KÜNNETH GEO

This clear and elegant text introduces Künneth, or l
foundations up, beginning with a rapid introduction .. ..,...... geometry at a level
suitable for undergraduate students. Unlike other books on this topic, it includes a
systematic development of the foundations of Lagrangian foliations.

The latter half of the text discusses Künneth geometry from the point of view of
basic differential topology, featuring both new expositions of standard material and
new material that has not previously appeared in book form. This subject, which has
many interesting uses and applications in physics, is developed ab initio, without
assuming any previous knowledge of pseudo-Riemannian or para-complex geometry.

This book will serve both as a reference work for researchers and as an invitation
for graduate students to explore this field, with open problems included as inspiration
for future research.

M.J.D. HAMILTON has been teaching mathematics at all levels at
Ludwig-Maximilians-Universität München and Universität Stuttgart for the past 15
years. He is interested in the interactions of geometry and theoretical physics, and is
known for his successful transfer of ideas between the two subjects. He is the author of
the acclaimed textbook *Mathematical Gauge Theory* (2017).

D. KOTSCHICK has been Professor of Mathematics, holding the Chair of
Differential Geometry, at Ludwig-Maximilians-Universität München for 25 years. A
researcher of exceptionally broad knowledge and interests, he is an internationally
recognised expert in several areas of geometry and topology. He is known for the
depth and insight of his research as well as for his meticulous scholarship and the
clarity of his writing. He is a long-standing member of the London Mathematical
Society and in 1996 received the Lucien Godeaux Prize of the Royal Society of Liège.

LONDON MATHEMATICAL SOCIETY LECTURE NOTE SERIES

Managing Editor: Professor Endre Süli, Mathematical Institute, University of Oxford, Woodstock Road, Oxford OX2 6GG, United Kingdom

The titles below are available from booksellers, or from Cambridge University Press at www.cambridge.org/mathematics

London Mathematical Society Student Texts

Künneth Geometry

Symplectic Manifolds and their Lagrangian Foliations

M. J. D. HAMILTON
Universität Stuttgart

D. KOTSCHICK
Ludwig-Maximilians-Universität München

CAMBRIDGE
UNIVERSITY PRESS

Shaftesbury Road, Cambridge CB2 8EA, United Kingdom

One Liberty Plaza, 20th Floor, New York, NY 10006, USA

477 Williamstown Road, Port Melbourne, VIC 3207, Australia

314–321, 3rd Floor, Plot 3, Splendor Forum, Jasola District Centre,
New Delhi – 110025, India

103 Penang Road, #05–06/07, Visioncrest Commercial, Singapore 238467

Cambridge University Press is part of Cambridge University Press & Assessment,
a department of the University of Cambridge.

We share the University's mission to contribute to society through the pursuit of
education, learning and research at the highest international levels of excellence.

www.cambridge.org
Information on this title: www.cambridge.org/9781108830713

DOI: 10.1017/9781108902977

First published 2024

A catalogue record for this publication is available from the British Library.

Library of Congress Cataloging-in-Publication Data
Names: Hamilton, Mark J. D., author. | Kotschick, Dieter, 1963- author.
Title: Künneth geometry : symplectic manifolds and their Lagrangian
foliations / M. J. D. Hamilton, D. Kotschick.
Description: Cambridge ; New York, NY : Cambridge University Press, 2024. |
Series: London Mathematical Society lecture note series 489 | Includes
bibliographical references and index.
Identifiers: LCCN 2023023076 | ISBN 9781108830713 (hardback)
Subjects: LCSH: Symplectic geometry. | Symplectic manifolds. | Foliations
(Mathematics)
Classification: LCC QA665 .H36 2024 | DDC 516.3/6–dc23/eng/20230809
LC record available at https://lccn.loc.gov/2023023076

ISBN 978-1-108-83071-3 Hardback
ISBN 978-1-108-82875-8 Paperback

Contents

1

Introduction

This book is about the geometry and topology of symplectic manifolds carrying pairs of complementary Lagrangian foliations. The resulting structure is very rich, intertwining symplectic geometry, the theory of foliations, dynamical systems, and pseudo-Riemannian geometry in interesting ways.

Before describing the contents of the book in detail, we would like to discuss a few motivating vignettes. The first two of these are to be kept in mind as motivational background, whereas the third and fourth ones will be taken up again and again later in the book.

1.1 Motivation

1.1.1 Pairs of Complementary Foliations

Let M be a smooth manifold, and $E \subset TM$ a smooth subbundle of the tangent bundle. We say that E is integrable if through every point $p \in M$ there is a local submanifold L_p with the property that for all $q \in L_p$ the tangent space $T_q L_p$ agrees with E_q. In particular, the dimension of L_p equals the rank of E. Such a submanifold is an integral manifold (of maximal dimension) for E. If E is integrable, then the maximal connected integral manifolds are the leaves of a foliation $\mathcal{F}$ with $T\mathcal{F} = E$. Any foliation is locally trivial in the sense that, in a suitable chart around any point, intersections of the leaves with the domain of the chart look like parallel affine subspaces in $\mathbb{R}^n$, generalising the flowbox picture for one-dimensional foliations.

Now assume that we have a foliation $\mathcal{F}$ on M. The question of whether $\mathcal{F}$ admits a complementary foliation $\mathcal{G}$ is interesting, and often very difficult. The complementarity condition is that $TM = T\mathcal{F} \oplus T\mathcal{G}$, where we do not mean that TM is only abstractly isomorphic to the Whitney sum on the right-hand

1

side, but the more stringent condition that the bundles on the right really are subbundles that form complements of each other inside TM at every point in M.

We can always choose a complement F for the subbundle $E = T\mathcal{F} \subset TM$, but in general an arbitrary complement is not integrable. Only when $\mathcal{F}$ has codimension 1, which means that the complement has rank 1, is it always integrable because all line fields are. Therefore the first interesting case to look at is that of a one-dimensional foliation on a three-manifold. In this case one looks for a two-dimensional foliation that is complementary to a given line field.

Consider the Hopf fibration $\pi \colon S^3 \longrightarrow S^2$, and the one-dimensional foliation $\mathcal{F}$ whose leaves are the fibres of π. In this case there is no complementary foliation. For if $\mathcal{G}$ were complementary to $\mathcal{F}$, then every leaf of $\mathcal{G}$ would be a connected covering space of S^2, and therefore diffeomorphic to S^2. We would then conclude that S^3 is diffeomorphic to $S^2 \times S^1$.

By the same argument, the foliation $\mathcal{F}$ whose leaves are the fibres of the non-trivial S^2-bundle over S^2 does not have a complementary foliation.

In the language of G-structures, a splitting $TM = E \oplus F$ into the direct sum of complementary subbundles of ranks p and q is a G-structure for the group $G = GL_p(\mathbb{R}) \times GL_q(\mathbb{R}) \subset GL_n(\mathbb{R})$, where n is the dimension of M. The question about the existence of such a splitting can often be answered in terms of algebraic topological invariants of M. Such a G-structure is integrable if and only if it is induced from a bifoliation, that is, a local product structure given by a pair of complementary foliations.

The question of the integrability of G-structures has been around since at least the 1950s. For example, it was raised by Calabi as Problem 9 in Hirzebruch's celebrated problem list [Hi-54]. We refer to [Hi-87] and [Kot-13] for accounts of what is now known about those problems.

For the particular type of G-structure at hand, if one does not require full integrability, but requires only the weaker condition that one of the two distributions is integrable, then a lot is known, since one is just asking for the existence of a foliation, of dimension p say, on M, assuming that the tangent bundle of M admits a rank p subbundle. In many cases the integrability of all distributions up to homotopy has been proved by Thurston, for example if $p = n - 1$ (see [Thu-76a]), and also if $p = 2$ (see [Thu-74]). In other cases there are additional obstructions coming from the Bott Vanishing Theorem that forces the vanishing of certain characteristic classes of the normal bundle of a foliation.

Returning to the full integrability of $GL_p(\mathbb{R}) \times GL_q(\mathbb{R})$-structures, it is still a very difficult problem to understand when a splitting of the tangent bundle can be induced by a bifoliation. As far as we know, the technology of h-principles emanating from Thurston's work does not apply in this case. Even in situations where both distributions are separately homotopic to integrable ones, it is very unclear whether they can simultaneously be made integrable in such a way that they remain complementary. It is certainly not possible to fix one foliation and then homotope the normal bundle to obtain a second, complementary, foliation. This problem appears already for $p = 1$ and $q = 2$, since there are circle bundles over surfaces which do not admit any horizontal foliation complementary to the fibres, as in the example of the Hopf fibration above. Although the two-dimensional horizontal subbundle is homotopic to a foliation, that foliation will never be complementary to the fibres. Of course in this case one can just switch the rôles of the two distributions and argue that one makes the two-dimensional distribution integrable without worrying about the complement since every one-dimensional distribution is integrable. This switching does not work even for $p = q = 2$. In this case all distributions are homotopic to integrable ones [Thu-74], but if we take for M^4 the non-trivial S^2-bundle over S^2, then again there is no two-dimensional foliation complementary to the fibres of the fibration. If one just homotopes the horizontal distribution to make it integrable (and no longer horizontal), then one does not know whether an integrable complement exists for the homotoped distribution.

In the case of a four-manifold whose tangent bundle splits as a Whitney sum of two rank 2 bundles, Thurston's Theorem [Thu-74] can be applied to each of the two subbundles to obtain two foliations. However, it is unknown whether one can always keep them complementary while making them both integrable. For example, it is an open problem whether $S^2 \widetilde{\times} S^2$ admits a pair of complementary two-dimensional foliations.

For general surface bundles over surfaces the existence or non-existence of a horizontal foliation is an interesting problem that has attracted quite a bit of attention in recent years, but is still open. We refer the interested reader to [KM-05, Bow-11, BCS-13] for discussions of this problem.

The existence of a bifoliation with $p = q$ is a special situation, which appears, for example, in the paper of Harvey and Lawson [HL-12], where it is called a double manifold, or a $\mathbb{D}$-manifold. On a surface this kind of structure can exist only if the Euler characteristic vanishes, but even in dimension 4 there are lots of examples. An example with non-zero Euler characteristic is given by the product of two surfaces of non-zero Euler characteristic. For vanishing Euler characteristic one can take the product of an arbitrary three-manifold with the circle. The existence of two-dimensional foliations on three-manifolds together with the integrability of line fields shows that every such four-manifold is a 'double manifold', but in general the foliations have complicated dynamics, and have little to do with the global product structure.

1.1.2 Hamiltonian Dynamics

The phase spaces M considered in classical mechanics are symplectic manifolds, so there is a symplectic form ω, which is a closed non-degenerate 2-form. Non-degeneracy means that the map

$$\mathfrak{X}(M) \longrightarrow \Omega^1(M)$$

$$X \longmapsto i_X\omega$$

given by contraction is an isomorphism between vector fields and 1-forms. Therefore, for any Hamiltonian function $H\colon M \longrightarrow \mathbb{R}$ there is a unique vector field $X_H \in \mathfrak{X}(M)$ defined by the equation $i_{X_H}\omega = dH$. The Hamiltonian dynamical system corresponding to H is the flow φ_t of the Hamiltonian vector field X_H. Using that ω is closed, we have

$$L_{X_H}\omega = i_{X_H}d\omega + di_{X_H}\omega = 0 + d^2H = 0 \,,$$

by the Cartan formula, which implies that $\varphi_t^*\omega = \omega$, so the Hamiltonian flow is a flow by symplectomorphisms.

To understand the dynamics of the system, it is useful to find conserved quantities, or first integrals. Using that ω is skew-symmetric, we calculate

$$L_{X_H}H = i_{X_H}dH = \omega(X_H, X_H) = 0 \,,$$

so H is always conserved under the flow, which is therefore along the level sets of H. For any function $f \in C^\infty(M)$ the condition that f be preserved under the flow φ_t is $df(X_H) = 0$, which can be rewritten as $\omega(X_f, X_H) = 0$. This motivates the definition of the Poisson bracket

$$\{f, g\} = \omega(X_f, X_g)$$

for any pair of smooth functions on M. If this vanishes, one says that the two functions are in involution. In this case the formula

$$i_{[X,Y]}\omega = L_X i_Y\omega - i_Y L_X\omega$$

shows that the corresponding vector fields X_f and X_g commute.

Since phase space is even-dimensional, say of dimension $2n$, the nicest possible situation is when there are n conserved quantities that are independent in a suitable sense. So let $H = f_1, \ldots, f_n$ be n conserved quantities for the

Hamiltonian flow φ_t, and assume that they are pairwise in involution, so all their Poisson brackets vanish. We consider the map

$$F: M \longrightarrow \mathbb{R}^n$$

$$x \longmapsto (f_1(x), \dots, f_n(x)) \,.$$

If c is a regular value of F, then the level set $M_c = F^{-1}(c)$ is an n-dimensional smooth submanifold of M. The condition that c is regular for F means that at every point in M_c the one-forms $df_1, \dots, df_n$ are linearly independent, and therefore the corresponding vector fields $X_{f_1}, \dots, X_{f_n}$ are also linearly independent. However, these vector fields are all tangent to M_c, and they commute. So M_c has a locally free $\mathbb{R}^n$-action. If M_c is compact and connected, it follows that it is a torus T^n. Moreover, on M_c all the contractions $i_{X_{f_i}}\omega$ vanish since the f_i are constant, and since the X_{f_i} span the tangent spaces to M_c at all points, we conclude that the restriction $\omega|M_c$ vanishes identically. Thus M_c is an example of a Lagrangian submanifold in a symplectic manifold.

If we look at the open set of M consisting of the regular points of F, this subset carries a foliation by Lagrangian submanifolds which are the individual level sets. This is the prototypical example of a Lagrangian foliation. A global perspective on this situation was discussed by Duistermaat [Dul-80], among others.

1.1.3 Anosov Symplectomorphisms

We now consider certain special discrete dynamical systems, which exhibit hyperbolic behaviour everywhere. They will be discussed in more detail in Subsection 5.3.1 of Chapter 5.

A diffeomorphism $f: M \to M$ of a compact manifold is Anosov if there is a continuous splitting of the tangent bundle into invariant subbundles of positive rank $TM = E^s \oplus E^u$ such that for all $k > 0$

$$\|Df^k(v)\| \le a \cdot e^{-bk}\|v\| \quad \forall v \in E^s \,,$$

$$\|Df^k(v)\| \ge a \cdot e^{bk}\|v\| \quad \forall v \in E^u \,,$$

for some positive constants a and b. Here the norms are taken with respect to some arbitrary Riemannian metric g. While the precise values of the constants a and b depend on the choice of g, the property of being Anosov does not. If the defining inequalities hold for some g, then they hold for every g (with different constants).

The defining property of an Anosov diffeomorphism is sometimes referred to as the existence of an Anosov splitting $TM = E^s \oplus E^u$ into stable (or contracting) and unstable (or dilating) subbundles E^s and E^u respectively. This means that f is hyperbolic everywhere. It is easy to see that when an Anosov splitting exists, it is uniquely determined by f, as the contracting and dilating subspaces have to be maximal with these properties.

The subbundles E^s and E^u are actually tangent to foliations of M with smooth leaves, although the distributions are only assumed continuous. The resulting foliations are called the stable and unstable foliations of f.

Suppose now that M is closed and symplectic, and f is an Anosov diffeomorphism which preserves the symplectic form, $f^*\omega = \omega$, so f is an Anosov symplectomorphism. Then E^s and E^u are Lagrangian with respect to ω, and therefore are tangent to a pair of complementary Lagrangian foliations.

To see this, suppose $v, w \in E^s$. Then

$$\omega(v, w) = (f^*\omega)(v, w) = \omega(Df(v), Df(w)) = \ldots = \omega(Df^k(v), Df^n(w)) \, .$$

Using the auxiliary metric g, we find that there is a constant c such that

$$|\omega(v, w)| \leq c \cdot \|\omega\| \cdot \|Df^k(v)\| \cdot \|Df^k(w)\| \leq c \cdot \|\omega\| \cdot a^2 \cdot e^{-2bk} \cdot \|v\| \cdot \|w\| \, .$$

Letting k go to infinity, the right-hand side becomes arbitrarily small. Therefore $\omega(v, w) = 0$, and E^s is ω-isotropic. By the same argument with f^{-1} replacing f we conclude that E^u is also ω-isotropic. As the two distributions are complementary, they must be equidimensional and Lagrangian.

We have seen that an Anosov symplectomorphism of M induces a Lagrangian bifoliation, and so one would naturally like to know how common this situation is. Even without the assumption that f preserves a symplectic form, the mere existence of an Anosov diffeomorphism seems to be a very strong assumption on M, and most manifolds should not admit any such diffeomorphism. The earliest problems and conjectures to this effect go back to Anosov and Smale. For example, Smale [Sma-67, Problem (3.5)] asked whether a closed manifold admitting an Anosov diffeomorphism must be covered by Euclidean space. This would be even stronger than just saying that M must be aspherical, a conclusion which is also still unknown. We refer the interested reader to [GL-16] for a recent discussion of the status of this problem.

In the situation of an Anosov symplectomorphism much more can be said. First of all, since the top-degree power of a symplectic form is a volume form, such diffeomorphisms are volume-preserving, and, in particular, topologically transitive. Second of all, an Anosov symplectomorphism preserves its associated Lagrangian bifoliation, and so is an automorphism of this structure. We will see in Chapter 5 that the automorphism group is in fact a Lie group. More generally, the pseudogroup of structure-preserving local diffeomorphisms of a bi-Lagrangian structure is a Lie pseudogroup, and bi-Lagrangian structures are rigid structures in the terminology of Gromov [Gro-88, DG-91]. This can be seen as the starting point for the work of Benoist and Labourie [BL-93], who proved that an Anosov symplectomorphism of a compact manifold M with smooth stable and unstable foliations is smoothly conjugate to a hyperbolic infranil automorphism. In particular, the manifold M has a nilpotent Lie group for its universal covering, answering Smale's question affirmatively. The proof of [BL-93] relies on the fact that Anosov symplectomorphisms are topologically transitive and act by automorphisms of a rigid structure, so that one can apply Gromov's open orbit theorem [Gro-88].

The result of [BL-93] is part of a long line of investigations which show that much more can be proved for Anosov diffeomorphisms with smooth stable and unstable foliations than for arbitrary Anosov diffeomorphisms, for which the Anosov splitting usually has very little regularity.

1.1.4 Affinely Flat Manifolds

A manifold M is called affinely flat if its (co-)tangent bundle admits a flat torsion-free connection. Equivalently, M has an atlas whose transition maps are affine transformations between open sets in Euclidean space, which, particularly in this case, should really be thought of as affine space.

For any connection on a vector bundle $E \longrightarrow M$, the horizontal subbundle is integrable if and only if the connection is flat. If E admits a flat connection, then M, embedded in E as the image of the zero-section, is a leaf of the horizontal foliation. The horizontal foliation together with the vertical foliation, whose leaves are the fibres of E, make up a bifoliation on the total space of E. If the rank of E equals the dimension of the base manifold M, then we have a bifoliation with equidimensional foliations, or what is called a double manifold in [HL-12].

If $E = T^*M$ happens to be the cotangent bundle of M, then the total space of this bundle has a tautological exact symplectic form, for which the fibres are Lagrangian. The condition for the horizontal foliation defined by a flat connection to be Lagrangian turns out to be precisely the torsion-freeness of the flat connection. This shows that the cotangent bundle of an affinely flat manifold carries a pair of complementary Lagrangian foliations. Moreover, M is a leaf of one of the two foliations, namely the horizontal one. We will give the details of these arguments in Section 4.2 of Chapter 4. These results are due to Weinstein [Wei-71], who proved them as a converse to his observation that the leaves of Lagrangian foliations are affinely flat with respect to the Bott connection. He thus obtained the characterisation of affinely flat manifolds as the manifolds that occur as leaves of Lagrangian foliations.

In spite of this, by now classical, characterisation of affinely flat manifolds in terms of Lagrangian foliations, symplectic geometry and the theory of foliations have so far not been used to address the many open problems about affinely flat manifolds. For example, there is a long-standing conjecture, usually attributed to Chern, suggesting that the Euler characteristics of closed affinely flat manifolds must vanish. This has been proved in many special cases; for example, Klingler [Kli-17] resolved the case of affinely flat manifolds with a parallel volume form. However, the general case of Chern's conjecture is still open, and one might hope that the theory of Lagrangian bifoliations might provide some insight into it.

1.2 What is in This Book?

We have seen that symplectic manifolds with pairs of complementary Lagrangian foliations arise naturally in various parts of geometry. It is this bi-Lagrangian structure we investigate in this book, studying its geometry, and also the topology of manifolds admitting such a structure. For reasons explained in Section 5.4 of Chapter 5, we call a symplectic structure together with a Lagrangian bifoliation a Künneth structure. In this book we set out the basics of Künneth geometry starting from symplectic geometry and the theory of foliations. We think of these considerations as *a priori* a part of differential topology. It turns out that there is an essentially canonical pseudo-Riemannian metric of neutral signature associated to a Künneth structure, but this arises *a posteriori* and is not part of our definition. When discussing this metric we do not assume that the reader has any expertise in pseudo-Riemannian geometry. Instead, along the way we explain how to adapt standard arguments in Riemannian geometry to the pseudo-Riemannian setting. We use only a little complex geometry, and no para-complex geometry at all, except to show that para-Kähler structures are in fact the same as Künneth structures.

In **Chapter 2** we discuss linear symplectic geometry, including the Linear Darboux Theorem, the space of Lagrangian subspaces, and bi-Lagrangian splittings. First we carry out this discussion in a single symplectic vector space and then extend it to symplectic vector bundles. We introduce linear Künneth structures, which are just bi-Lagrangian splittings of symplectic vector bundles. The existence of a Künneth structure on a vector bundle turns out to impose strong restrictions on its characteristic classes.

Chapter 3 constitutes a quick introduction to symplectic manifolds and their Lagrangian submanifolds. We introduce the Moser homotopy method, and use it to prove the Darboux Theorem for symplectic forms and Weinstein's Tubular Neighbourhood Theorem for Lagrangian submanifolds.

In **Chapter 4** we give a brief introduction to foliations and flat bundles. We relate the integrability of subbundles of the tangent bundle to both the flatness and the torsion-freeness of certain affine connections. We also extend this discussion to almost product structures, obtaining criteria for the integrability of such a structure to a bifoliation. We introduce Lagrangian foliations of symplectic manifolds, and we discuss the Bott connection, first for general foliations, and then in more detail for Lagrangian foliations. We also adapt the Moser argument from Chapter 3 to prove the Darboux Theorem for a symplectic form together with a Lagrangian foliation.

In **Chapter 5** we begin the development of Künneth geometry itself. We give the basic definitions, and we note that instead of the usual Darboux Theorem one can prove a local normal form statement that involves a function that plays the rôle of the Kähler potential in Kähler geometry. We also introduce not necessarily integrable almost Künneth structures, which are linear Künneth structures on the tangent bundles of manifolds. We explain why every almost Künneth structure has a natural pseudo-Riemannian metric, making it into a rigid geometric structure with a small automorphism group. Most of this chapter is taken up with constructions of examples. Our emphasis is on global constructions yielding examples of Künneth structures on closed manifolds. Some of the examples we obtain have not appeared in the literature before now.

In **Chapter 6** we prove that every almost Künneth structure gives rise to a preferred affine connection for which the structure is parallel. This connection is torsion free if and only if the structure is integrable to a Künneth structure. In the integrable case only, the Künneth connection is the Levi–Civita connection of the associated pseudo-Riemannian metric. Moreover, its restriction to the two Lagrangian foliations equals the respective Bott connection. At the end of this chapter we prove the equivalence between Künneth and para-Kähler structures.

In **Chapter 7** we investigate the curvature of the Künneth connection. Some of this is done for arbitrary almost Künneth structures, but after the initial discussion of the general case we soon restrict to integrable structures, for which more can be said. We prove that a standard Darboux theorem holds for a Künneth structure if and only if the curvature vanishes. Unfortunately this does not yield a uniformisation result, because the Künneth connection is usually not complete. We work out explicit formulas for the Ricci and scalar curvatures. The formula for the Ricci curvature gives a criterion for when the neutral pseudo-Riemannian metric associated with a Künneth structure is an Einstein metric. At the end of this chapter we investigate parallel Künneth structures on Kähler manifolds.

In **Chapter 8** we introduce hypersymplectic structures. In keeping with our discussion of Künneth structures, we give a purely symplectic formulation that does not involve a pseudo-Riemannian metric or a connection as part of the definition. We do, however, show that our definition is equivalent to the usual metric definition. We show that every hypersymplectic structure gives rise to a family of Künneth structures parametrised by the circle. The leaves of the corresponding Lagrangian foliations are not just affinely flat, which is true for all Lagrangian foliations, but are also symplectic, equipped with symplectic forms that are parallel with respect to the flat affine connection. At the end of this chapter we prove that hypersymplectic structures, equivalently, their subordinate Künneth structures, are Ricci-flat, or neutral Calabi–Yau.

Chapter 9 contains a quick introduction to nil- and infra-nilmanifolds. This is motivated by the fact that Anosov symplectomorphisms can exist only on infra-nilmanifolds [BL-93]. More generally, nilmanifolds offer the possibility of reducing the construction of geometric structures to linear algebra by passing back and forth between left-invariant structures on a Lie group and the corresponding linear structures on its Lie algebra. We use this approach to give explicit examples of Anosov symplectomorphisms and of hypersymplectic – and therefore Künneth – structures on nilmanifolds. We also classify left-invariant Künneth structures on four-dimensional nilpotent Lie groups. At the end of this chapter we indicate how to generalise to solvmanifolds in place of nilmanifolds.

In **Chapter 10** we investigate (almost) Künneth structures on closed four-manifolds. After a brief introduction to the classical invariants of closed smooth four-manifolds, we use these invariants to characterise those closed four-manifolds that admit an almost Künneth structure. In particular, we prove that the existence of an almost Künneth structure does not constrain the fundamental group. We then show that the candidates for having an integrable Künneth structure are the symplectic Calabi–Yau manifolds, whose topology is very restricted. In particular, their fundamental groups are very special. The known examples of symplectic Calabi–Yau manifolds of real dimension 4 are, up to finite coverings, the $K3$ manifold and T^2-bundles over T^2. For the latter we make a systematic study of Lagrangian foliations and of Künneth structures. Many of the results in this chapter are new.

Chapter 10 freely uses results about the Seiberg–Witten invariants of (symplectic) four-manifolds. We do not explain those results, but quote them as needed, giving precise references. Those results are not used elsewhere in this book, and treating them fully would require us to write a completely different book.

1.3 How to Read This Book

Throughout this book we assume familiarity with the basic language of smooth manifolds. No specialised knowledge of differential or symplectic geometry is required.

The first five chapters can be used as a textbook for a rapid introduction to symplectic geometry and the study of Lagrangian foliations aimed at undergraduates. For this audience one could leave out the parts of Chapter 2 that discuss characteristic classes. A course covering these five chapters would fit neatly into a term with only eight or nine weeks of lectures.

In a longer course for beginning graduate students, with twelve to fifteen weeks of lectures, one can cover most of the book. For this audience one can probably cover Chapters 1 to 3 quite quickly, then treat Chapters 4 to 8 in considerable detail, and finally switch to a survey mode for Chapters 9 and 10. Indeed the original manuscript for this book was formed by the lecture notes of such a course that we taught at the University of Munich in the spring semester of 2016.

1.4 What is *Not* in This Book

The only curvature conditions we discuss for Künneth structures are flatness, leading to the best possible local normal form, and the Einstein condition. The latter is satisfied, for example, for the Künneth structures arising from hyper-symplectic structures. There are many other curvature conditions one could consider, and that have been considered in the literature, but that we do not discuss here, and those lead to many special results in (local) pseudo-Riemannian geometry.

We do not discuss homogeneous Künneth structures, except on nilmanifolds. It was proved by Hou, Deng and Kaneyuki [HDK-97] that a manifold with a Künneth structure homogeneous under a compact Lie group must be a torus. For structures homogeneous under a non-compact semisimple Lie group, Hou, Deng, Kaneyuki and Nishiyama [HDKN-99] proved that the manifold must be an adjoint orbit of a hyperbolic element. This means that in both these cases one does not find any interesting compact examples. We refer to the survey by Alekseevsky, Medori and Tomassini [AMT-09] for further information on the homogeneous situation.

We also do not discuss calibrations and special Lagrangian submanifolds in Künneth geometry, but refer the reader to the paper by Harvey and Lawson [HL-12] and the references therein. As explained in those references, calibrations in Künneth geometry appear naturally in the classical Monge–Kantorovich mass transport problem.

There have been many other instances in which Künneth structures have arisen in connection with various differential equations. As noted by Hitchin [Hit-90] when he first introduced hypersymplectic structures, these structures arise naturally on moduli spaces of harmonic maps from Riemann surfaces to compact Lie groups, and on moduli spaces of solutions for the KdV equation and for the non-linear Schrödinger equation. The common feature of these equations that leads to the connection with hypersymplectic geometry is that they are dimensional reductions of the self-dual Yang–Mills equations in signature $(2, 2)$. Very recently, a variant of Nahm's equations was added to this list, again making contact with hypersymplectic geometry; see Bielawski, Romão and Röser, [BiRR-17]. Another class of differential equations, this time arising from hydrodynamics, was connected to hypersymplectic geometry by Banos, Roubtsov and Roulstone [BaRR-16].

Quite recently Künneth structures have found applications in Teichmüller theory; see Loustau and Sanders [LS-17]. This is perhaps not surprising given the appearance of Künneth vector bundles in the guise of symplectic Anosov structures in the work of Burger, Iozzi, Labourie and Wienhard [BILW-05].

Acknowledgement

We are grateful to V. Guillemin for inspiration, to the members of the audience of our course at the Ludwig-Maximilians-Universität München for their helpful comments and suggestions, and to R. Coelho, A. Jackson and G. Placini for help with proofreading.

2

Linear Algebra and Bundle Theory

In this chapter we discuss the linear algebra of symplectic vector spaces and symplectic vector bundles. To prepare the ground for the discussion of Künneth structures on manifolds in later chapters we introduce linear Künneth structures on vector bundles, and we work out consequences of the existence of Künneth structures in terms of characteristic classes.

The earlier parts of this chapter contain standard material that some readers may be able to skip. There is a substantial overlap, for example, with Chapter 2 of the book of McDuff–Salamon [McS-95]. The later parts contain some important results that are used throughout the book. While not original, these results clarify some of the folklore around symplectic vector bundles and their Lagrangian subbundles. Our reference for the theory of characteristic classes is Milnor–Stasheff [MS-74].

2.1 Linear Algebra

2.1.1 Linear Symplectic Forms

Here is the most basic definition, which is the beginning of all of symplectic mathematics.

Definition 2.1 Let V be a finite-dimensional (real) vector space. A *symplectic form* on V is a 2-form $\omega \in \Lambda^2 V^*$ that is *non-degenerate* in the sense that

$$\omega(v, u) = 0 \quad \text{for all } u \in V$$

implies $v = 0$. Equivalently,

$$\omega(v, V) \equiv 0 \Rightarrow v = 0$$

or

$$i_v\omega = \omega(v, -) \equiv 0 \Rightarrow v = 0 \,.$$

We call (V, ω) a *symplectic vector space*.

One can characterise non-degenerate two-forms in the following way.

Lemma 2.2   *A two-form $\omega \in \Lambda^2 V^*$ is non-degenerate if and only if the map*

$$\phi_\omega : V \longrightarrow V^*$$
$$v \longmapsto i_v\omega = \omega(v, -)$$

is an isomorphism.

Proof   In one direction, since V and V^* have the same dimension, the linear map ϕ_ω is an isomorphism if and only if it is injective. Conversely, injectivity of ϕ_ω is equivalent to the non-degeneracy of ω.   □

Example 2.3   Let $V = \mathbb{R}^{2n}$ with basis

$$(e_1, \ldots, e_n, f_1, \ldots, f_n) \,.$$

We denote the dual basis of V^* by

$$(\alpha_1, \ldots, \alpha_n, \beta_1, \ldots, \beta_n) \,.$$

Then

$$\omega = \sum_{i=1}^{n} \alpha_i \wedge \beta_i$$

is a symplectic form on V. It is uniquely characterised by

$$\omega(e_i, f_j) = \delta_{ij} \quad \text{(Kronecker delta)} \,,$$
$$\omega(e_i, e_j) = 0 = \omega(f_i, f_j) \,.$$

The map ϕ_ω is given by

$$\phi_\omega(e_i) = \beta_i \,,$$
$$\phi_\omega(f_i) = -\alpha_i \,, \quad \forall i = 1, \ldots, n,$$

showing that it is indeed an isomorphism between V and V^*.

Just as in the case of scalar products, there is a notion of a symplectic orthogonal for a subspace.

Definition 2.4 Let (V, ω) be a symplectic vector space and $U \subset V$ a linear subspace. The *symplectic orthogonal* of U is defined as

$$U^{\perp\omega} = \{v \in V \mid \omega(v, U) \equiv 0\}$$
$$= \{v \in V \mid \omega(v, u) = 0 \quad \forall u \in U\}.$$

In other words, if $i\colon U \hookrightarrow V$ is the injection, then $U^{\perp\omega}$ is the kernel of the linear map

$$V \xrightarrow{\phi_\omega} V^* \xrightarrow{i^*} U^*.$$

We will now prove that, up to a choice of basis, every linear symplectic form has the form given in Example 2.3. This is often called the *Linear Darboux Theorem*, because it is the infinitesimal version of the Darboux Theorem for symplectic forms on manifolds, to be proved later.

Theorem 2.5 (Linear Darboux Theorem) *Let ω be a symplectic form on a vector space V. Then there exists a basis*

$$(e_1, \ldots, e_n, f_1, \ldots, f_n)$$

of V with dual basis

$$(\alpha_1, \ldots, \alpha_n, \beta_1, \ldots, \beta_n)$$

of V^* such that ω is given by

$$\omega = \sum_{i=1}^{n} \alpha_i \wedge \beta_i.$$

Such a basis of V (or V^*) is called a *symplectic basis* with respect to ω.

Proof Since ω is non-degenerate, it is not identically zero, so there exist vectors $e_1, f_1 \in V$ with $\omega(e_1, f_1) = 1$. We set

$$V_1 = \operatorname{span}\{e_1, f_1\}.$$

Since ω is non-degenerate on V_1, the symplectic orthogonal $V_1^{\perp\omega}$ intersects V_1 only in the zero vector, and is a complement to V_1; compare Lemma 2.10 below. We claim that the restriction $\omega|_{V_1^{\perp\omega}}$ is non-degenerate. For the proof suppose there exists a vector v in $V_1^{\perp\omega}$ with

$$\omega\left(v, V_1^{\perp\omega}\right) = 0.$$

Since we also have

$$\omega(v, V_1) = 0,$$

and V_1 and $V_1^{\perp\omega}$ are complementary, this would give

$$\omega(v, V) = 0 .$$

By the non-degeneracy of ω on V we conclude $v = 0$.

We can now find a symplectic basis for V by induction on the dimension, replacing V by $V_1^{\perp\omega}$ in the inductive step. $\qquad\square$

Corollary 2.6 *If ω is a symplectic form on a real vector space V, then the dimension of V is even, $\dim V = 2n$.*

Corollary 2.7 *A two-form ω on a vector space V of dimension $2n$ is symplectic if and only if*

$$\omega^n = \underbrace{\omega \wedge \cdots \wedge \omega}_{n} \in \Lambda^{2n}V^*$$

is non-zero, i.e. a volume form on V. In particular, every symplectic vector space has a canonical orientation defined by ω^n.

Proof If ω is symplectic, we can choose a symplectic basis for V and calculate ω^n. We then see that $\omega^n \neq 0$. Conversely, assume that ω is not symplectic, so that there exists a non-zero vector $v \in V$ with $i_v\omega = 0$. Then also $i_v(\omega^n) = 0$, and ω^n is not a volume form. $\qquad\square$

Structure-preserving maps of symplectic vector spaces are called symplectomorphisms.

Definition 2.8 Let (V, ω_V) and (W, ω_W) be symplectic vector spaces. A linear isomorphism $f \colon V \to W$ is called a *symplectomorphism* if

$$f^*\omega_W = \omega_V .$$

If such an f exists, then (V, ω_V) and (W, ω_W) are called *symplectomorphic*.

We can rephrase the Linear Darboux Theorem (Theorem 2.5) to state that all symplectic vector spaces of the same dimension are symplectomorphic to one another.

2.1.2 Subspaces in Symplectic Vector Spaces

Let (V, ω) be a symplectic vector space of dimension $2n$. We are interested in linear subspaces of V that are in a special position with respect to ω.

Definition 2.9 Let $U \subset V$ be a linear subspace.

(i) We call U *symplectic* if the restriction $\omega|_U$ is symplectic.

(ii) We call U *isotropic* if the restriction $\omega|_U$ vanishes identically.

(iii) We call U *Lagrangian* if it is isotropic and

$$\dim U = \frac{1}{2}\dim V = n.$$

In terms of the symplectic orthogonal, one has the following.

Lemma 2.10 *Let $U \subset V$ be a linear subspace. The following hold:*

(i) $\dim U + \dim U^{\perp\omega} = \dim V$,

(ii) $(U^{\perp\omega})^{\perp\omega} = U$,

(iii) U *is symplectic if and only if* $U \cap U^{\perp\omega} = 0$,

(iv) U *is isotropic if and only if* $U \subset U^{\perp\omega}$,

(v) *if U is isotropic, then* $\dim U \leq \frac{1}{2}\dim V$.

Proof Since $U^{\perp\omega}$ is the kernel of

$$V \xrightarrow{\psi_\omega} V^* \xrightarrow{i^*} U^* ,$$

ϕ_ω is an isomorphism and i^* is an epimorphism, we have

$$\dim U^{\perp\omega} = \dim V - \dim U^* .$$

This proves the first claim. For the second, we first prove $U \subset (U^{\perp\omega})^{\perp\omega}$. Fix $u \in U$. Then

$$\omega(u, v) = 0 \quad \forall v \in U^{\perp\omega} ,$$

hence $u \in (U^{\perp\omega})^{\perp\omega}$. Using part (i), this inclusion cannot be strict, and thus (ii) holds.

For the third claim, we have $U \cap U^{\perp\omega} \neq 0$ if and only if there exists a $v \in U$ such that $\omega(v, U) = 0$. But this happens if and only if $\omega|_U$ is not symplectic.

For the fourth claim, we have $U \subset U^{\perp\omega}$ if and only if $\omega(U, U) \equiv 0$, i.e. if and only if U is isotropic.

Finally, if U is isotropic, then by (i) and (iv) we have

$$\dim V = \dim U + \dim U^{\perp\omega} \geq 2\dim U ,$$

proving the fifth claim. $\qquad\square$

Corollary 2.11 *A linear subspace U in a symplectic vector space (V, ω) is Lagrangian if and only if $U = U^{\perp\omega}$. A Lagrangian subspace is an isotropic subspace of maximal dimension.*

Example 2.12   Let (V, ω) be a symplectic vector space of dimension $2n$. Choose a symplectic basis

$$(e_1, \ldots, e_n, f_1, \ldots, f_n)$$

for V. Then

$$L = \mathrm{span}(e_1, \ldots, e_n)$$

and

$$L' = \mathrm{span}(f_1, \ldots, f_n)$$

are Lagrangian subspaces. Note that these subspaces are complementary, i.e. $L \oplus L' = V$.

Conversely we have the following result.

Proposition 2.13   *Let (V, ω) be a symplectic vector space of dimension $2n$ and $L \subset V$ a Lagrangian subspace. Any basis $(e_1, \ldots, e_n)$ for L can be completed to a symplectic basis $(e_1, \ldots, e_n, f_1, \ldots, f_n)$ of V.*

Proof   Consider the span $U = \mathrm{span}\{e_2, \ldots, e_n\}$. The symplectic orthogonal $U^{\perp_\omega}$ has dimension $n + 1$ and contains L, which has dimension n. Hence there exists a vector $f_1 \in U^{\perp_\omega}$ that is not an element of L. It satisfies

$$\omega(e_i, f_1) = 0 \quad \forall i = 2, \ldots, n$$

but

$$\omega(e_1, f_1) \neq 0 \,,$$

since f_1 is not an element of L and L is maximally isotropic. Normalising f_1 we can assume that

$$\omega(e_1, f_1) = 1 \,.$$

As in the proof of Theorem 2.5, let

$$V_1 = \mathrm{span}\{e_1, f_1\} \,.$$

Then ω restricts to a symplectic form on the orthogonal $V_1^{\perp_\omega}$ and U is a Lagrangian subspace in this symplectic vector space. By induction, we construct a symplectic basis

$$(e_1, \ldots, e_n, f_1, \ldots, f_n)$$

for V. □

Corollary 2.14   *Let $L \subset V$ be a Lagrangian subspace in a symplectic vector space. Then there exists a complementary Lagrangian subspace L' with $V = L \oplus L'$.*

We also have the following Linear Darboux Theorem for a pair (V, L) consisting of a symplectic vector space and a Lagrangian subspace.

Corollary 2.15 *Let (V, ω_V) and (W, ω_W) be symplectic vector spaces of the same dimension and assume that $L_V \subset V$ and $L_W \subset W$ are Lagrangian subspaces. Then there exists a symplectomorphism $f : V \to W$ taking L_V onto L_W.*

We can adapt the proof of Proposition 2.13 to show the following:

Proposition 2.16 *Let (V, ω) be a symplectic vector space of dimension $2n$ and $L, L' \subset V$ two complementary Lagrangian subspaces. For any basis $(e_1, \ldots, e_n)$ of L there exists a* unique *basis $(f_1, \ldots, f_n)$ of L', so that*

$$(e_1, \ldots, e_n, f_1, \ldots, f_n)$$

is a symplectic basis of V.

Proof In the proof of Proposition 2.13 the subspace $U^{\perp\omega}$ intersects L' in a one-dimensional subspace. Hence there is a unique vector f_1 in this intersection with

$$\omega(e_1, f_1) = 1.$$

By induction, this implies the claim. □

2.1.3 The Space of Lagrangian Complements

Let (V, ω) be a symplectic vector space. In Corollary 2.14 we showed that every Lagrangian subspace $L \subset V$ has a Lagrangian complement L', so that $L \oplus L' = V$. However, the Lagrangian complement L' is not unique. For the applications to Lagrangian distributions on manifolds, it is important to understand the space of all Lagrangian complements to a given Lagrangian subspace L.

We will approach this issue using the following construction of a tautological symplectic vector space. Let W be an arbitrary real vector space of dimension n, and $V = W \oplus W^*$. Elements of V consist of pairs $(w, \lambda) \in W \oplus W^*$. We define a map

$$\omega_0 : V \times V \longrightarrow \mathbb{R}$$

by

$$\omega_0((w_1, \lambda_1), (w_2, \lambda_2)) = \lambda_1(w_2) - \lambda_2(w_1).$$

It is clear that ω_0 is a skew-symmetric two-form on V. Moreover, it is non-degenerate and therefore symplectic. To see this, note that the map

$$\phi_{\omega_0} : W \oplus W^* \longrightarrow W^* \oplus W$$

is given by

$$(w, \lambda) \longmapsto (\lambda, -w) ,$$

which is an isomorphism.

It is clear that W and W^* are Lagrangian subspaces of (V, ω_0). Corollary 2.15 tells us that, for any pair consisting of a symplectic vector space V of dimension $2n$ and a Lagrangian subspace W, there exists a symplectomorphism onto the pair (V, W). To understand the space of Lagrangian complements for a Lagrangian in a general symplectic vector space it therefore suffices to understand the space of Lagrangian complements to W in (V, ω_0).

Lemma 2.17 *A vector subspace $A \subset V$ is complementary to W if and only if $\pi_2|_A : A \to W^*$ is an isomorphism. If A is complementary to W, then A is the graph of a uniquely determined linear map $\alpha : W^* \to W$.*

Proof The subspace A is complementary to W if and only if $A \cap W = 0$ and A has dimension n. The first fact is equivalent to $\pi_2|_A$ being injective. This implies the first claim.

Suppose A is complementary to W. Then the inverse of $\pi_2|_A$ is of the form

$$(\pi_2|_A)^{-1} : W^* \longrightarrow A \subset W \oplus W^*$$
$$\lambda \longmapsto \quad (\alpha(\lambda), \lambda) .$$

Since the inverse of $\pi_2|_A$ is linear, the map α itself is linear. This proves the second claim. $\square$

We want to understand when a complementary subspace A is Lagrangian. For this we use the natural pairing

$$\langle \cdot, \cdot \rangle : W \times W^* \longrightarrow \mathbb{R}$$
$$(w, \lambda) \longmapsto \lambda(w) .$$

A linear map $\alpha : W^* \to W$ is *self-adjoint* with respect to this pairing if

$$\langle \alpha(\mu), \lambda \rangle = \langle \alpha(\lambda), \mu \rangle \quad \forall \mu, \lambda \in W^* .$$

We then have the following result.

Proposition 2.18 *Let $A \subset V = W \oplus W^*$ be a linear subspace complementary to W, defined as the graph of a linear map $\alpha : W^* \to W$. Then A is Lagrangian with respect to ω_0 if and only if α is self-adjoint.*

Proof The subspace A is Lagrangian if and only if ω_0 vanishes on all pairs of vectors $(\alpha(\lambda), \lambda)$ and $(\alpha(\mu), \mu)$ in V. For the evaluation we have

$$\lambda(\alpha(\mu)) - \mu(\alpha(\lambda)) = \langle \alpha(\mu), \lambda \rangle - \langle \alpha(\lambda), \mu \rangle \,.$$

This vanishes if and only if α is self-adjoint. □

Since there is a one-to-one correspondence between complements A and linear maps $\alpha \colon W^* \to W$, the space of all Lagrangian complements to W can be identified with the vector space of symmetric, real $(n \times n)$-matrices.

We summarise this discussion in the following result.

Theorem 2.19 *Let (V, ω) be an arbitrary symplectic vector space of dimension $2n$ and $L \subset V$ a Lagrangian subspace. Then the space $\mathcal{L}(V, \omega, L)$ of all Lagrangian subspaces $L' \subset V$ complementary to L is a real vector space of dimension $\frac{1}{2}n(n+1)$.*

2.1.4 Compatible Complex Structures

Recall that a complex structure on a real vector space V is an isomorphism $J \colon V \to V$ such that $J^2 = -\mathrm{Id}_V$. This makes V into a complex vector space by declaring scalar multiplication by $i \in \mathbb{C}$ to be the application of J.

If V is a symplectic vector space, there are compatibility conditions one can impose on the symplectic and complex structures.

Definition 2.20 Let (V, ω) be a symplectic vector space. A complex structure $J \colon V \to V$ is *tamed* by the symplectic form if

$$\omega(v, Jv) > 0 \quad \forall v \neq 0 \in V \,.$$

A complex structure J on V is called *compatible* with the symplectic form ω if

$$g_J(v, w) = \omega(v, Jw)$$

defines a positive-definite scalar product g_J on V.

We denote by $\mathcal{J}(V, \omega)$ the space of all complex structures on V compatible with the given symplectic form ω.

Lemma 2.21 *A complex structure J on V is compatible with a symplectic form ω if and only if it is tamed by ω and ω is J-invariant in the sense that*

$$\omega(Jv, Jw) = \omega(v, w) \quad \forall v, w \in V \,.$$

Proof The complex structure is compatible with the symplectic form ω if and only if g_J is symmetric and positive definite, i.e.

$$g_J(v, w) = g_J(w, v) \quad \forall v, w \in V ,$$
$$g_J(v, v) > 0 \quad \forall v \neq 0 \in V .$$

The second condition is equivalent to J being tamed by ω and the first condition is equivalent to

$$\omega(Jv, Jw) = \omega(v, w) \quad \forall v, w \in V .$$

$\square$

Proposition 2.22 *Let (V, ω) be a symplectic vector space. Then there exists a compatible complex structure J.*

Proof According to the Linear Darboux Theorem (Theorem 2.5) there exists a symplectic basis

$$(e_1, \ldots, e_n, f_1, \ldots, f_n)$$

for V. Then $J: V \to V$, defined by

$$Je_i = f_i ,$$
$$Jf_i = -e_i ,$$

is a compatible complex structure. $\square$

We want to understand the space $\mathcal{J}(V, \omega)$ of all compatible complex structures. In particular, we want to show that this space is contractible. There are several ways to prove this; the way we do it here involves Lagrangian subspaces.

Lemma 2.23 *Let (V, ω) be a symplectic vector space of dimension $2n$ with a compatible complex structure J. If $L \subset V$ is a Lagrangian subspace then JL is a complementary g_J-orthogonal Lagrangian subspace to L.*

Proof Since J is an isomorphism, the subspace JL has dimension n. It is Lagrangian, because

$$\omega(Jv, Jw) = \omega(v, w) = 0 \quad \forall v, w \in L .$$

Furthermore, it is g_J-orthogonal to L, for if $v \in L$ and $w \in JL$, then $Jw \in L$ and

$$g_J(v, w) = \omega(v, Jw) = 0 .$$

$\square$

Proposition 2.24 *Let (V, ω) be a symplectic vector space and $L \subset V$ a Lagrangian subspace. For every positive-definite scalar product h on L and every Lagrangian complement L' there exists a unique, ω-compatible complex structure J on V such that $L' = JL$ and $g_J|_L = h$.*

Proof We first prove the existence of J. To do so we choose an h-orthonormal basis $(e_1, \ldots, e_n)$ for L and let $(f_1, \ldots, f_n)$ be the unique basis for L' so that both bases together form a symplectic basis for V; see Proposition 2.16. We then define a complex structure J on V by

$$Je_i = f_i \,,$$
$$Jf_i = -e_i \,.$$

By definition, we have $JL = L'$. It is clear that J is compatible with ω, because the basis $(e_1, \ldots, e_n, f_1, \ldots, f_n)$ is symplectic. In addition, $(e_1, \ldots, e_n)$ is an orthonormal basis of L for both h and $g_J|_L$, hence $g_J|_L = h$.

To show uniqueness, suppose that J and J' are two complex structures that satisfy the condition in the proposition. Choose again an h-orthonormal basis $(e_1, \ldots, e_n)$ for L, and let

$$f_j = Je_j \,,$$
$$f'_j = J'e_j \,.$$

The vectors $\{f_j\}$ and $\{f'_j\}$ each form a basis for L'. It suffices to show that

$$f_j = f'_j$$

for all indices j, since then $J \equiv J'$ on all of V.

By the uniqueness statement of Proposition 2.16 it suffices to show that

$$(e_1, \ldots, e_n, f_1, \ldots, f_n)$$

and

$$(e_1, \ldots, e_n, f'_1, \ldots, f'_n)$$

are symplectic bases for V. We show this for the first basis; the argument is the same for the second basis. We have

$$\omega(e_i, e_j) = 0 = \omega(f_i, f_j) \,,$$

since L and L' are Lagrangian. Furthermore,

$$\omega(e_i, f_j) = \omega(e_i, Je_j) = g_J(e_i, e_j) = h(e_i, e_j) = \delta_{ij} \,,$$

since J is ω-compatible. This proves the claim. $\qquad\square$

We can now describe the space of compatible complex structures.

Theorem 2.25 *Let (V, ω) be a symplectic vector space. Fix some Lagrangian subspace L and denote by $\mathrm{Met}(L)$ the space of all positive-definite scalar products on L. Then the map*

$$F \colon \mathcal{J}(V, \omega) \longrightarrow \mathcal{L}(V, \omega, L) \times \mathrm{Met}(L)$$
$$J \longmapsto \quad\quad (JL, g_J|_L)$$

is a bijection, and a homeomorphism with respect to the natural topologies.

In particular, the space $\mathcal{J}(V, \omega)$ of complex structures compatible with ω is contractible.

Proof The map F is well defined and continuous. In Proposition 2.24 we constructed the continuous inverse.

By Theorem 2.19 the space $\mathcal{L}(V, \omega, L)$ is a vector space, and therefore is contractible. The space of metrics $\mathrm{Met}(L)$ is not a vector space, but is convex, and therefore contractible as well. □

To end this subsection, we need to discuss the relationship between orientations and Lagrangian splittings of symplectic vector spaces. Recall that a symplectic vector space (V, ω) of dimension $2n$ has a canonical orientation defined by ω^n. This can also be thought of as the orientation defined by a compatible complex structure J.

Suppose that we are given a decomposition $V = L \oplus L'$ into complementary Lagrangian subspaces. Then it is possible to choose J so that it maps L isomorphically to L', and so any Lagrangian splitting has the form $V = L \oplus L$.

Lemma 2.26 *Let (V, ω) be a symplectic vector space of dimension $2n$ and $V = L \oplus L$ a splitting into complementary Lagrangian subspaces. Choose an orientation of L. Then the product orientation on $L \oplus L$ differs from the symplectic orientation of V by the sign*

$$\epsilon(n) = (-1)^{\frac{n(n-1)}{2}}.$$

Proof We can choose a symplectic basis

$$(e_1, \ldots, e_n, f_1, \ldots, f_n)$$

for V, where the e_i are an oriented basis for L, and the f_i are an oriented basis for $L' = L$. The symplectic orientation of V corresponds to

$$e_1 \wedge f_1 \wedge \ldots \wedge e_n \wedge f_n$$

and we need

$$1 + 2 + 3 + 4 + \cdots + (n-1) = \frac{1}{2}n(n-1)$$

many transpositions to change

$$e_1 \wedge e_2 \wedge \ldots \wedge e_n \wedge f_1 \wedge f_2 \wedge \ldots \wedge f_n$$

to the symplectic orientation. This proves the claim.   □

2.2 Symplectic and Complex Vector Bundles

Let $\pi \colon E \longrightarrow M$ be a smooth real vector bundle over a smooth manifold M.

Definition 2.27 A *symplectic structure* on the vector bundle E is a smooth section $\omega \in \Gamma(\Lambda^2 E^*)$ with the property that, for all $x \in M$, the form $\omega(x)$ is symplectic on the fibre E_x.

In other words, ω is a fibrewise symplectic form on the fibres of E varying smoothly in the neighbourhood of any point. We will sometimes refer to the pair (E, ω) as a symplectic vector bundle.

Since symplectic vector spaces are even-dimensional and canonically oriented by the top power of the symplectic form, every symplectic vector bundle is oriented and of even rank $2n$.

Definition 2.28 A *complex structure* on the vector bundle E is a smooth section $J \in \Gamma(\mathrm{End}(E))$ with the property that, for all $x \in M$, the endomorphism $J(x)$ is a complex structure on the fibre E_x.

Such a J makes E into a complex vector bundle, with scalar multiplication by i being given by the application of J.

It turns out that every symplectic vector bundle is a complex vector bundle in an essentially unique way. To discuss this we use compatibility between symplectic forms and complex structures, which can be formulated for bundles in the same way as we did for single vector spaces.

Theorem 2.29 *Let (E, ω) be a symplectic vector bundle. Then there exists a compatible complex structure J on E. Moreover, this complex structure is unique up to homotopy.*

Proof Let $\mathcal{J}(E, \omega) \to M$ be the locally trivial fibre bundle associated to E whose fibre over x consists of the space $\mathcal{J}(E_x, \omega(x))$ of compatible complex structures. According to Theorem 2.25, the fibres of this bundle are contractible. Therefore the bundle admits a section, and, moreover, any two sections are homotopic through sections.   □

Conversely, if we are given a complex structure J on an arbitrary real vector bundle E, we can always choose a J-invariant positive-definite fibre-wise scalar product g. Then

$$\omega(v, w) = g(Jv, w)$$

is skew-symmetric. Since J is invertible and g is non-degenerate, it follows that ω is non-degenerate and therefore a symplectic structure on E. Moreover, J is compatible with this symplectic structure. The space of possible scalar products g in this construction is convex, and so up to homotopy ω is independent of g and depends only on J.

To summarise, symplectic and complex vector bundles are really the same (up to suitable notions of equivalence). We want to extend this equivalence to include Lagrangian subbundles on the symplectic side.

Definition 2.30 Let (E, ω) be a symplectic vector bundle over a smooth manifold M. A subbundle $L \subset E$ is called *Lagrangian* if the fibre L_x is Lagrangian in the symplectic vector space $(E_x, \omega(x))$, for all $x \in M$.

Proposition 2.31 *Let (E, ω) be a symplectic vector bundle over a smooth manifold M, and $L \subset E$ a Lagrangian subbundle. Then there exists a complementary Lagrangian subbundle L', so that $L \oplus L' = E$.*

Proof Let $\mathcal{L}(E, \omega, L) \to M$ be the smooth fibre bundle associated to E and L whose fibre over $x \in M$ consists of the space $\mathcal{L}(E_x, \omega(x), L_x)$ of Lagrangian subspaces in E_x complementary to L_x. Since the fibres of this bundle are contractible according to Theorem 2.19, the bundle has a global section L' over M. $\square$

With Proposition 2.24 we obtain the following.

Theorem 2.32 *Let (E, ω) be a symplectic vector bundle of rank $2n$ over a smooth manifold M. Suppose $E = L \oplus L'$ is a splitting of E into two complementary Lagrangian subbundles. Choose an arbitrary, positive-definite bundle metric h on L. Then there exists a unique complex structure J on E, compatible with ω, such that $L' = JL$ and $g_J|_L = h$.*

Recall that a totally real subbundle $F \subset E$ in a complex vector bundle (E, J) is a subbundle with the property that $J(F_x) \cap F_x = 0$ for all $x \in M$. By the following result, totally real subbundles of maximal rank correspond to Lagrangian subbundles in symplectic vector bundles.

Theorem 2.33 *A symplectic vector bundle (E, ω) of rank $2n$ admits a Lagrangian subbundle $L \subset E$ if and only if the corresponding complex vector bundle (E, J) admits a totally real subbundle F of rank n.*

Proof Given a Lagrangian subbundle L, the previous theorem gives a complex structure J for which L is totally real. Conversely, given a totally real subbundle F in (E, J), of maximal rank n, we can choose a J-invariant metric g so that F and $J(F)$ are g-orthogonal. Then the symplectic structure ω defined by $\omega(v, w) = g(Jv, w)$ on E has F as a Lagrangian subbundle. □

Since a complex structure $J\colon E \to E$ is an orientation-preserving bundle isomorphism, Lemma 2.26 implies the following.

Corollary 2.34 *Let (E, ω) be a symplectic vector bundle of rank $2n$ over a smooth manifold M. Suppose $E = L \oplus L'$ is a splitting of E into two complementary Lagrangian subbundles.*

(i) *The vector bundles L and L' are isomorphic as real,* unoriented *vector bundles over M.*
(ii) *If L is orientable and we fix an orientation, then the product orientation on $L \oplus L$ differs from the symplectic orientation of E by the sign*

$$\epsilon(n) = (-1)^{\frac{n(n-1)}{2}} .$$

2.3 Künneth Vector Bundles

We can now define linear Künneth or bi-Lagrangian structures on vector bundles.

Definition 2.35 A *Künneth vector bundle* is a symplectic vector bundle (E, ω) together with a splitting $E = L \oplus L'$ into complementary Lagrangian subbundles.

Note that an ω-compatible J can be chosen so that it gives an isomorphism between L and L', and so the Lagrangian splitting always has the form $L \oplus L$. We will sometimes refer to the triple (E, ω, L) as a (linear) *Künneth structure*.

The definition of a Künneth structure can be reformulated in several ways.

Proposition 2.36 *Let (E, ω) be a symplectic vector bundle of rank $2n$. The following conditions are equivalent:*

(i) *(E, ω) admits a Künneth structure,*
(ii) *(E, ω) admits a Lagrangian subbundle $L \subset E$,*
(iii) *the corresponding complex vector bundle (E, J) admits a totally real subbundle of rank n,*

(iv) *the corresponding complex vector bundle (E, J) is isomorphic to the com-plexification $L \otimes_{\mathbb{R}} \mathbb{C}$ of a real vector bundle L of rank n.*

Proof Clearly the first condition implies the second. Moreover, the existence of Lagrangian complements in Proposition 2.31 gives the converse.

The second and third conditions are equivalent by Theorem 2.33.

The third condition implies the fourth since if $F \subset E$ is totally real for J, then (E, J) is $\mathbb{C}$-linearly isomorphic to $F \otimes_{\mathbb{R}} \mathbb{C}$. Conversely, if (E, J) is isomorphic to $L \otimes_{\mathbb{R}} \mathbb{C}$, then $E = L \oplus iL$, and both summands are totally real. □

A Künneth vector bundle is in particular symplectic and therefore orientable and oriented. However, the Lagrangian subbundle L may very well be non-orientable. This motivates the following definition.

Definition 2.37 A Künneth structure (E, ω, L) is *orientable* if L is an orientable vector bundle.

The existence of a Künneth structure on a vector bundle will impose restric-tions on its characteristic classes. As usual, we will call Chern classes of a symplectic vector bundle (E, ω) the Chern classes of the corresponding com-plex vector bundle (E, J). Since J is unique up to homotopy, the Chern classes are independent of the exact choice we make for J.

Theorem 2.38 *Let (E, ω) be a symplectic vector bundle admitting a La-grangian subbundle $L \subset E$. Then the odd-degree Chern classes $c_{2i+1}(E) \in H^{4i+2}(M; \mathbb{Z})$ are two-torsion classes. If the Lagrangian subbundle L is ori-entable, then $c_1(E) = 0$.*

Proof Under the assumption of the theorem, the complex vector bundle (E, J) is the complexification of L, and its underlying real bundle is isomorphic to $L \oplus L$. It follows that

$$c_{2i+1}(E) = \beta(w_{2i}(L) \cup w_{2i+1}(L)) , \qquad (2.1)$$

where the w_j denote the Stiefel–Whitney classes and

$$\beta \colon H^{4i+1}(M; \mathbb{Z}_2) \longrightarrow H^{4i+2}(M; \mathbb{Z})$$

is the Bockstein homomorphism associated to multiplication by 2 in the coef-ficients. See for example Problem 15-D in [MS-74].

If L is orientable, so that $w_1(L) = 0$, then we obtain

$$c_1(E) = \beta(w_1(L)) = 0 \in H^2(M; \mathbb{Z}) . \qquad (2.2)$$

□

This is of course a strong obstruction to the existence of orientable Künneth structures, because most real vector bundles do not admit a complex structure with vanishing first Chern class.

Another obstruction comes from the Euler class:

Corollary 2.39 *Let E be a Künneth vector bundle of real rank $4k + 2$. Then the Euler class $e(E)$ is a two-torsion class.*

Proof For any complex vector bundle, the Euler class of the underlying oriented real vector bundle equals the top Chern class. By (2.1) this is a 2-torsion class, since we assumed that the complex rank of E was odd. □

If (E, ω) is a Künneth bundle with Lagrangian subbundle L, then, as a real unoriented vector bundle, E is isomorphic to $L \oplus L$. If L is orientable and oriented, then $L \oplus L$ has a product orientation induced from that of L. By Corollary 2.34 this agrees with the canonical, symplectic or complex, orientation of E after multiplication by

$$\epsilon(n) = (-1)^{\frac{n(n-1)}{2}} ,$$

where n is the complex rank of E, which is the real rank of L. Thus the isomorphism $E \cong L \oplus L$ is orientation-preserving if $\epsilon(n) = 1$, and orientation-reversing otherwise. If $\epsilon(n) = -1$, we can think of the oriented bundle E as $L \oplus \overline{L}$, where $\overline{L}$ denotes L endowed with the reversed orientation.

Now, by the Whitney sum formula for the Euler class, we have

$$e(E) = e(L) \cup \epsilon(n)e(L) = \epsilon(n)e(L)^2 . \tag{2.3}$$

If n is odd, then $e(L)$ is a two-torsion class, and so is $e(E)$, which is what we saw above. However, when n is even, $e(L)^2$ may well be non-torsion.

Notes for Chapter 2

1. Using a different terminology, Künneth vector bundles were considered by Bejan in [Bej-93].
2. Dazord [Daz-81] claimed that the (real) Euler class of any Künneth bundle vanishes. His argument was that the Chern–Weil integrand vanishes identically if one chooses an orthogonal connection adapted to the splitting $E = L \oplus L$. A moment's thought about permutations shows that this argument requires the same dimension assumption as Proposition 2.39, when the claim reduces to (2.1). In the Erratum [Daz-85], Dazord mentions the dimension assumption, and then goes on to claim that $e(E_{\mathbb{R}}) = e(L)^2$, missing the sign in (2.3). This sign will be crucial in our considerations of tangent bundles of four-manifolds in Chapter 10.

3

Symplectic Geometry

This chapter is a crash course on symplectic geometry. We first introduce symplectic manifolds and their Lagrangian submanifolds and give some examples. Then we discuss the Moser method [Mos-65] and use it to prove results about the existence of local normal forms and of standard tubular neighbourhoods. This is of course just the beginning of the vast area of symplectic geometry. A more extensive account is contained, for example, in the book of McDuff–Salamon [McS-95]; see especially Chapter 3 of that book. Other useful references for this material include the writings of Weinstein [Wei-71, Wei-77].

3.1 Symplectic Forms on Manifolds

Definition 3.1 Let M be a smooth manifold. A two-form $\omega \in \Omega^2(M)$ is called *non-degenerate* or *almost symplectic* if ω_p is non-degenerate on the tangent space T_pM for all $p \in M$. The two-form ω is called *symplectic* if it is non-degenerate and closed, i.e. $d\omega = 0$.

If ω is a non-degenerate two-form on a manifold M it follows that the dimension of M is even, $\dim M = 2n$, and

$$\omega^n \in \Omega^{2n}(M)$$

is a volume form; in particular, M has a canonical orientation, which we refer to as the symplectic orientation.

If ω is symplectic, then $d\omega = 0$, hence ω represents a cohomology class

$$[\omega] \in H^2_{dR}(M) \cong H^2(M; \mathbb{R}) .$$

If the manifold M is closed, i.e. compact without boundary, and we denote by

$[M] \in H_{2n}(M; \mathbb{Z})$ the fundamental class determined by the symplectic orientation, then

$$\langle [\omega]^n, [M] \rangle = \langle [\omega] \cup \cdots \cup [\omega], [M] \rangle > 0 .$$

In particular we have the following result.

Proposition 3.2 *Let (M, ω) be a closed symplectic manifold of dimension $2n$. Then the cup products*

$$[\omega]^k = \underbrace{[\omega] \cup \cdots \cup [\omega]}_{k \ times} \in H^{2k}(M; \mathbb{R}) , \quad \forall k \in \{1, \ldots, n\} ,$$

are non-zero. In particular, all even Betti numbers $b_{2k}(M)$ for $1 \le k \le n$ are non-zero.

As in the linear case, structure-preserving maps are called symplectomorphisms.

Definition 3.3 A diffeomorphism $\phi \colon M \to N$ between symplectic manifolds (M, ω_M) and (N, ω_N) is called a *symplectomorphism* if

$$\phi^* \omega_N = \omega_M .$$

Example 3.4 (i) Every symplectic vector space is a symplectic manifold. In particular, if $\mathbb{R}^{2n}$ has coordinates $(x_1, \ldots, x_n, y_1, \ldots, y_n)$, then

$$\omega_0 = \sum_{i=1}^{n} dx_i \wedge dy_i$$

is a symplectic form on $\mathbb{R}^{2n}$.

(ii) Let Σ be an orientable 2-dimensional manifold (a surface) and let ω be a volume form. Then ω is non-degenerate and $d\omega = 0$, because all forms of degree greater than 2 vanish on Σ. Hence (Σ, ω) is a symplectic manifold.

(iii) Let S^m be an m-dimensional sphere. Because of Proposition 3.2 and the previous example the manifold S^m has a symplectic form if and only if $m = 2$.

(iv) All Kähler manifolds are symplectic, because the Kähler form is a symplectic form. In particular, complex projective space $\mathbb{CP}^n$ and all smooth projective complex-algebraic varieties are symplectic manifolds.

3.1.1 Products and Fibre Bundles

It is obvious that if (M^{2m}, ω_M) and (N^{2n}, ω_N) are symplectic manifolds, then the product manifold $M \times N$ has a symplectic form given by $\omega = \pi_M^* \omega_M + \pi_N^* \omega_N$.

Example 3.5   Let $(\Sigma_1, \omega_1), \ldots, (\Sigma_k, \omega_k)$ be oriented surfaces. Then

$$M = \Sigma_1 \times \cdots \times \Sigma_k$$

is a $2k$-dimensional symplectic manifold. In particular,

$$M^4 = \Sigma_g \times \Sigma_h$$

is a symplectic four-manifold, where Σ_g and Σ_h denote the closed, oriented surfaces of genus g and h. In the same way, all even-dimensional tori T^{2n} are symplectic, because they are products of n copies of T^2. In fact, if

$$(x_1, \ldots, x_n, y_1, \ldots, y_n)$$

denote standard coordinates on $\mathbb{R}^{2n}$, then the symplectic form

$$\omega = \sum_{i=1}^{n} dx_i \wedge dy_i$$

is invariant under the action of $\mathbb{Z}^{2n}$ by translations and hence induces a well-defined symplectic form on the quotient $T^{2n} = \mathbb{R}^{2n}/\mathbb{Z}^{2n}$.

Instead of products one can, in certain cases, consider non-trivial fibre bundles. The simplest case is when the fibre is a surface.

Theorem 3.6 (Thurston [Thu-76b])   *Let*

$$\Sigma^2 \longrightarrow N^{2k+2}$$
$$\downarrow \pi$$
$$M^{2k}$$

be a orientable fibre bundle, where Σ is a closed, oriented surface and (M, ω_M) is a closed, symplectic manifold. Suppose that

$$[\Sigma] \neq 0 \in H_2(N; \mathbb{R}).$$

Then N admits a symplectic form.

The homological condition is always satisfied if Σ is not a torus.

Proof   We can show that the class $[\Sigma]$ is non-zero in $H_2(N; \mathbb{R})$ if Σ is not a torus. For this let $V \subset TN$ denote the vertical tangent bundle over N, consisting of the tangent spaces to the fibres. Let $e(V) \in H^2(N; \mathbb{Z})$ denote the Euler class of the oriented two-plane bundle V. Then

$$\langle e(V), [\Sigma] \rangle = \chi(\Sigma)$$

is the Euler characteristic of Σ. Hence if this Euler characteristic is non-zero, then $[\Sigma] \neq 0 \in H_2(N; \mathbb{R})$.

The idea of the proof of the theorem is to construct a closed two-form α on N such that α restricts to a volume form on each fibre of the bundle. This is where we use the assumption that the homology class represented by Σ is non-zero. Once α has been constructed, let

$$\omega_N = \alpha + r\pi^*\omega_M \,,$$

where $r > 0$ is some positive number. Then ω_N is closed,

$$d\omega_N = d\alpha + r\pi^* d\omega_M = 0$$

and

$$\omega_N^{k+1} = \sum_{l=1}^{k+1} \binom{k+1}{l} \alpha^l \wedge r^{k+1-l} \pi^*(\omega_M^{k+1-l}) \,.$$

If r is sufficiently large, then the largest term in this sum is

$$r^k \binom{k+1}{1} \alpha \wedge \pi^*(\omega_M^k) = r^k(k+1)\alpha \wedge \pi^*(\omega_M^k) \,,$$

which is a volume form on N. Therefore ω_N^{k+1} will also be a volume form and ω_N is symplectic. □

Corollary 3.7 *Let*

$$\Sigma_h \longrightarrow N^4$$
$$\downarrow \pi$$
$$\Sigma_g$$

be an orientable surface bundle over a surface, where g is arbitrary and h ≠ 1. Then N admits a symplectic form.

Remark 3.8 Concerning the existence of symplectic forms on orientable surface bundles over surfaces it only remains to consider the case of bundles with T^2-fibres. For T^2-bundles over S^2 it is known that only the trivial product $T^2 \times S^2$ admits a symplectic form, since all other bundles have $b_2 = 0$. An example of such a bundle is $S^1 \times S^3$, where $S^3 \to S^2$ is the Hopf fibration. Orientable T^2-bundles over T^2 will be considered in Section 10.6 of Chapter 10. Finally, an orientable T^2-bundle over Σ_g, for $g > 1$, admits a symplectic form if *and only if* the class of the fibre is non-zero in real homology [Walc-05].

3.1.2 Cotangent Bundles

Let M^n be a smooth n-dimensional manifold and consider the total space of the cotangent bundle T^*M, which is a smooth, $2n$-dimensional manifold. We want to show that T^*M admits a canonical symplectic form. This situation occurs in classical mechanics, where M is the configuration space of a mechanical system and T^*M is the associated phase space.

Denote by $\pi: T^*M \to M$ the projection with differential

$$\pi_*: T_\alpha T^*M \longrightarrow T_x M$$

whenever $\pi(\alpha) = x$. We first define a tautological one-form $\lambda \in \Omega^1(T^*M)$.

Definition 3.9 Let $\alpha \in T^*M$ and v be a tangent vector to T^*M at α. The *canonical* or *Liouville* one-form λ on the manifold T^*M is defined by

$$\lambda_\alpha(v) = \alpha\left(\pi_*(v)\right) .$$

This one-form has the following property.

Lemma 3.10 *If $\gamma \in \Omega^1(M)$ is any given one-form on M, i.e. a smooth section $\gamma: M \to T^*M$, then*

$$\gamma^*\lambda = \gamma .$$

Proof Let $x \in M$ and $w \in T_x M$. Then

$$(\gamma^*\lambda)_x(w) = \lambda_{\gamma(x)}(\gamma_*(w)) = \gamma_x\left(\pi_*\gamma_*(w)\right) = \gamma_x(w) .$$

This proves the claim. $\square$

Let $(x_1, \ldots, x_n)$ be local coordinates in a chart $U \subset M$ and write a one-form $\alpha \in T_x^*M$ as $\alpha = \sum_{i=1}^n y_i dx_i$. We set

$$\bar{x}_i = x_i \circ \pi .$$

Then $(\bar{x}_1, \ldots, \bar{x}_n, y, \ldots, y_n)$ are local coordinates on T^*U.

Lemma 3.11 *In these coordinates on T^*U the Liouville one-form is given by*

$$\lambda = \sum_{i=1}^n y_i d\bar{x}_i .$$

Proof Let (x, α) be a point in T^*U with coordinates $(\bar{x}_1, \ldots, \bar{x}_n, y_1, \ldots, y_n)$. Then

$$\lambda_{(x,\alpha)}(\partial_{\bar{x}_i}) = \alpha\left(\pi_*(\partial_{\bar{x}_i})\right) = \left(\sum_{i=1}^n y_i dx_i\right)(\partial_{x_i}) = y_i$$

and

$$\lambda_{(x,\alpha)}(\partial_{y_i}) = \alpha\left(\pi_*(\partial_{y_i})\right) = \alpha(0) = 0 \ .$$

This proves the claim. $\qquad\square$

Theorem 3.12 *Let T^*M be the cotangent bundle of a smooth n-dimensional manifold M with Liouville one-form λ. Then*

$$\omega_{\mathrm{can}} = -d\lambda$$

is a symplectic form on T^*M.

In the literature one sometimes finds this definition with the opposite sign, so that $\omega_{\mathrm{can}} = d\lambda$.

Proof Since ω_{can} is an exact two-form, it is in particular closed. According to Lemma 3.11 we have in local coordinates $(\bar{x}_1, \ldots, \bar{x}_n, y_1, \ldots, y_n)$ on T^*U the formula

$$\omega_{\mathrm{can}} = \sum_{i=1}^{n} d\bar{x}_i \wedge dy_i,$$

showing that ω_{can} is non-degenerate. $\qquad\square$

3.1.3 Compatible Almost Complex and Hermitian Structures

An almost complex structure on a manifold M is a smooth section J of the endomorphism bundle $\mathrm{End}(TM)$ satisfying $J^2 = -\mathrm{Id}_{TM}$. Such a structure makes the tangent bundle TM into a complex vector bundle. As a corollary to Theorem 2.29 we have the following.

Corollary 3.13 *Let (M, ω) be an almost symplectic manifold. Then there exists a compatible almost complex structure J on M, unique up to homotopy. The corresponding metric g_J is a Riemannian metric on M.*

Definition 3.14 Let M be a smooth manifold. A triple (ω, J, g_J) of compatible almost symplectic, almost complex and Riemannian structures is called an *almost Hermitian structure*. It is called an *almost Kähler structure* if ω is closed, i.e. symplectic.

3.2 The Moser Method

In this section we explain and apply the so-called Moser method, originally invented by Moser [Mos-65] to study volume forms on closed manifolds.

Theorem 3.15 (Moser method)   *Let W be a 2n-dimensional manifold and $Q \subset W$ a compact submanifold. Assume that ω_0 and ω_1 are two closed two-forms on W that are identical and non-degenerate on the tangent space $T_q W$ for each $q \in Q$. Then there exist open neighbourhoods V_0 and V_1 of Q in W and a diffeomorphism*

$$\psi \colon V_0 \longrightarrow V_1 \, ,$$

isotopic to the identity, such that

$$\psi|_Q = \mathrm{Id}_Q \quad and \quad \psi^* \omega_1 = \omega_0 \, .$$

Proof   Let V be a tubular neighbourhood of Q. Then the difference

$$\tau = \omega_1 - \omega_0$$

is exact on V, so that $\tau = d\sigma$. To prove this, let $j \colon Q \to V$ be the inclusion. Since the tubular neighbourhood is diffeomorphic to the normal bundle νQ, which retracts onto the zero section Q, the map j induces an isomorphism

$$j^* \colon H^2_{dR}(V) \longrightarrow H^2_{dR}(Q) \, .$$

Since $j^* \omega_0 = j^* \omega_1$ by assumption, we have $j^*[\omega_0] = j^*[\omega_1]$ and hence $[\omega_0] = [\omega_1]$ in the cohomology of V. This implies the claim. In fact, using a direct construction of σ, one can check that we can assume that σ vanishes on $T_q W$ for all $q \in Q$.

We define a family of closed two-forms ω_t on W by

$$\omega_t = \omega_0 + t(\omega_1 - \omega_0) = \omega_0 + t d\sigma \, .$$

On the tangent bundle of W restricted to Q we have $\omega_t \equiv \omega_0$, which is non-degenerate. Hence if we choose the tubular neighbourhood V small enough, we can assume that ω_t is non-degenerate on V for all $t \in [0, 1]$.

We can define a unique vector field X_t on V by

$$i_{X_t} \omega_t = -\sigma.$$

Then $X_t \equiv 0$ along Q. Let ψ_t be its flow:

$$\frac{d}{dt}\psi_t = X_t \circ \psi_t \, .$$

If we choose a sufficiently small tubular neighbourhood $V_0 \subset V$, then the flow ψ_t is defined for all $x \in V_0$ and all $t \in [0, 1]$. The image $\psi_1(V_0) = V_1$ is an open neighbourhood of Q. We have $\psi_t|_Q = \mathrm{Id}_Q$ for all $t \in [0, 1]$.

Using the Cartan formula for the Lie derivative we find:

$$\frac{d}{dt}(\psi_t^*\omega_t) = \psi_t^*\left(\frac{d}{dt}\omega_t + L_{X_t}\omega_t\right)$$

$$= \psi_t^*(d\sigma + di_{X_t}\omega_t) = d\left(\psi_t^*(\sigma + i_{X_t}\omega_t)\right) = 0 .$$

Hence $\psi_t^*\omega_t$ is constant and

$$\psi_1^*\omega_1 = \psi_0^*\omega_0 = \omega_0 .$$

Setting $\psi = \psi_1$ this proves the claim. $\qquad\square$

An immediate corollary of the Moser method is the Darboux Theorem.

Theorem 3.16 (Darboux Theorem) *Let (M,ω) be a symplectic manifold of dimension $2n$ and $p \in M$ a point. Then there exists an open neighbourhood U of p with local coordinates $(x_1,\ldots,x_n,y_1,\ldots,y_n)$ such that the symplectic form is given by*

$$\omega|_U = \sum_{i=1}^{n} dx_i \wedge dy_i .$$

Proof We have to find an open neighbourhood U of p and a chart $\tau \colon U \to \mathbb{R}^{2n}$ such that

$$\tau^* \sum_{i=1}^{n} dx_i \wedge dy_i = \omega .$$

Consider the symplectic vector space (T_pM, ω_p). According to the Linear Darboux Theorem (Theorem 2.5) there exist linear coordinates $(x_1,\ldots,x_n)$ and $(y_1,\ldots,y_n)$ on T_pM such that

$$\omega_p = \sum_{i=1}^{n} dx_i \wedge dy_i .$$

Let $\phi \colon V \to V'$ be the diffeomorphism given by the exponential map of a Riemannian metric on M, where $V \subset T_pM$ is a small open neighbourhood of 0 and V' is an open neighbourhood of p in M. We define a symplectic form ω_1 on V' by

$$\phi^*\omega_1 = \omega_p = \sum_{i=1}^{n} dx_i \wedge dy_i .$$

We also have the symplectic form $\omega_0 = \omega$ on V'.

We now apply the Moser method of Theorem 3.15 for the compact submanifold $Q = \{p\} \subset M$. We have

$$\omega_0|_{T_pM} = \omega_p = \omega_1|_{T_pM},$$

since the differential of ϕ is the identity in $0 \in T_pM$. Hence there exists a diffeomorphism

$$\psi \colon V_0 \to V_1 \,,$$

where $V_0, V_1 \subset V'$ are open neighbourhoods of p, such that

$$\psi^*\omega_1 = \omega_0 \,.$$

Let $U = V_0$ and

$$\tau = \phi^{-1} \circ \psi \colon U \to T_pM \cong \mathbb{R}^{2n} \,.$$

Then

$$\tau^* \sum_{i=1}^{n} dx_1 \wedge dy_i = \psi^* \left(\phi^{-1}\right)^* \phi^*\omega_1 = \psi^*\omega_1 = \omega_0 = \omega \,.$$

This completes the proof. $\qquad\qquad\qquad\qquad\qquad\qquad\qquad\qquad$ □

3.3 Neighbourhoods of Lagrangian Submanifolds

In this section we discuss special submanifolds of symplectic manifolds, including Lagrangian submanifolds. From the Moser method we derive a normal form for the symplectic structure in a suitable tubular neighbourhood of a Lagrangian submanifold.

Definition 3.17 Let (M, ω) be a symplectic manifold. A submanifold $N \subset M$ is called *symplectic, isotropic* or *Lagrangian* if the tangent space T_pN has the corresponding property as a subspace of the symplectic vector space (T_pM, ω_p) for all points $p \in N$.

Here are some examples.

Example 3.18 Let (M, ω) and (N, η) be symplectic manifolds. On $V = M \times N$ we can consider the symplectic form

$$\omega_V = \pi_1^*\omega - \pi_2^*\eta \,.$$

Of course we could have taken the sum rather than the difference in the formula, but there is a reason for the choice of sign!

The two factors are symplectic submanifolds (for either sign), and the graph of a diffeomorphism $f \colon M \longrightarrow N$, if there is one, is Lagrangian if and only if f is a symplectomorphism, i.e. $f^*\eta = \omega$. (This is where the sign matters.)

Example 3.19   Let (T^*M, ω_{can}) be the cotangent bundle of a manifold M. Then the zero section of T^*M is a Lagrangian submanifold. The fibres are also Lagrangian submanifolds, and form the simplest example of a Lagrangian foliation; cf. Section 4.2 of Chapter 4 below.

We can generalise the previous example:

Proposition 3.20   *Let (T^*M, ω_{can}) be the cotangent bundle of a manifold M. Let $\alpha \colon M \to T^*M$ be a section, i.e. a one-form $\alpha \in \Omega^1(M)$. Then the image $L = \alpha(M)$ is a submanifold of T^*M, diffeomorphic to M, that is Lagrangian if and only if α is closed.*

Proof   Since $\alpha \colon M \to T^*M$ is a section, it is an embedding of M into T^*M. The image L is Lagrangian if and only if

$$0 = \alpha^* \omega_{can} = -\alpha^* d\lambda_{can} = -d\,(\alpha^* \lambda_{can}) = -d\alpha .$$

Here we used Lemma 3.10. $\square$

Example 3.21   Consider the torus $T^{2n} = \mathbb{R}^{2n}/\mathbb{Z}^{2n}$ with the symplectic form induced from the symplectic form

$$\omega = \sum_{i=1}^{n} dx_i \wedge dy_i$$

on $\mathbb{R}^{2n}$ with coordinates $(x_1, \ldots, x_n, y_1, \ldots, y_n)$. Let

$$\pi_1 \colon T^{2n} \longrightarrow T^n$$
$$[x_1, \ldots, x_n, y_1, \ldots, y_n] \longmapsto [x_1, \ldots, x_n]$$

be the projection onto the first n components. Then π_1 is a fibration whose fibres are Lagrangian tori. There exists a similar fibration π_2 onto the last n components, which is also Lagrangian. Hence T^{2n} has two complementary Lagrangian fibrations.

We saw in Example 3.19 that the zero section in a cotangent bundle of a manifold L is a Lagrangian submanifold, diffeomorphic to L. We now want to show that this situation is very generic: if L is a Lagrangian submanifold in some symplectic manifold M, then a tubular neighbourhood of L in M is symplectomorphic to a neighbourhood of the zero section in T^*L (even the fact that it is diffeomorphic to such a neighbourhood is non-trivial). The following is Weinstein's Lagrangian Tubular Neighbourhood Theorem.

Theorem 3.22 (Weinstein [Wei-71])   *Let (M, ω) be a symplectic manifold and $L \subset M$ a compact Lagrangian submanifold. Then there exists a tubular* *neighbourhood $V \subset M$ of L, a tubular neighbourhood $U \subset T^*L$ of the zero section and a diffeomorphism $\phi \colon U \to V$ so that*

$$\phi|_L = \mathrm{Id}_L \quad and \quad \phi^*\omega = \omega_{\mathrm{can}}.$$

Here we canonically identify L with the zero section in T^*L.

For the proof we will need the following lemma.

Lemma 3.23   *Let (M, ω) be a symplectic manifold and $L \subset M$ a compact Lagrangian submanifold.*

(i) *The normal bundle $\nu(L)$ of L in M is isomorphic to the tangent bundle TL, with the isomorphism given by any compatible almost complex structure J. The normal bundle is the orthogonal complement to the tangent bundle with respect to the Riemannian metric g_J, determined by ω and J.*

(ii) *The tangent bundle TL is isomorphic to the cotangent bundle T^*L. An isomorphism $\Phi \colon TL \to T^*L$ is determined by g_J:*

$$\Phi_x \colon T_xL \longrightarrow T_x^*L$$
$$v \longmapsto g_J(v, -).$$

Thus the bundles T^*L and $\nu(L)$ over L are isomorphic.

Proof   The first part is clear by Lemma 2.23. The second part is standard Riemannian geometry.   □

We can now prove Theorem 3.22.

Proof of Theorem 3.22   We apply the Moser method of Theorem 3.15. Note that here we have symplectic forms on two different manifolds, M and T^*L, while in the Moser method we assumed that the symplectic forms are defined on the same manifold. We will first construct a diffeomorphism

$$f \colon T^*L \supset U' \longrightarrow V' \subset M$$

between an open neighbourhood U' of the zero section in T^*L and an open tubular neighbourhood V' of the submanifold L in M. We then consider the symplectic forms $f^*\omega$ and ω_{can} on $U' \subset T^*L$ and apply the Moser method to these symplectic forms.

Recall from Riemannian geometry that the exponential map of a Riemannian metric on M defines an embedding

$$\exp \colon \nu(L) \supset W' \longrightarrow V' \subset M$$

onto an open tubular neighbourhood V' of L in M. Using the isomorphisms given by the lemma, we get a diffeomorphism

$$f \colon T^*L \supset U' \longrightarrow V' \subset M$$
$$(x, \alpha) \longmapsto \exp_x(J_x\Phi_x^{-1}(\alpha)) \,.$$

Both maps are defined on suitable open neighbourhoods W', U' of the zero sections in the corresponding bundles.

We claim that

$$f^*\omega = \omega_{\mathrm{can}}$$

on the tangent spaces of T^*L along the zero section. The theorem then follows by the Moser method.

We want to calculate the differential of f along the zero section in T^*L. We have

$$T_{(x,0)}T^*L = T_xL \oplus T_x^*L \,.$$

For a tangent vector

$$\mu = (v, \alpha) \in T_xL \oplus T_x^*L$$

we get

$$D_{(x,0)}f(\mu) = v + J_x\Phi_x^{-1}(\alpha) \,.$$

(This follows by calculating the differential on $(v, 0)$ and $(0, \alpha)$ separately.)

We can now calculate $f^*\omega$ along the zero section. Let

$$\mu = (v, \alpha) \,,$$
$$\eta = (w, \beta)$$

be tangent vectors in $T_{(x,0)}T^*L$. Then

$$\begin{aligned}
f^*\omega(\mu, \eta) &= \omega_x(v + J_x\Phi_x^{-1}(\alpha), w + J_x\Phi_x^{-1}(\beta)) \\
&= \omega_x(v, J_x\Phi_x^{-1}(\beta)) - \omega_x(w, J_x\Phi_x^{-1}(\alpha)) \\
&= g_J(v, \Phi_x^{-1}(\beta)) - g_J(w, \Phi_x^{-1}(\alpha)) \\
&= \beta(v) - \alpha(w) \,.
\end{aligned}$$

Here we used that T_xL and JT_xL are Lagrangian. On the other hand, we have

$$\begin{aligned}
\omega_{\mathrm{can}}(\mu, \eta) &= \left(\sum_{i=1}^n d\bar{x}_i \wedge dy_i\right)((v, \alpha), (w, \beta)) \\
&= \beta(v) - \alpha(w) \,.
\end{aligned}$$

This proves the claim. $\qquad\qquad\qquad\qquad\qquad\qquad\qquad\qquad\qquad\quad$ $\square$

4

Foliations and Connections

In this chapter we introduce foliations and discuss some fundamental examples. We characterise the integrability of subbundles of tangent bundles in terms of both flatness and torsion-freeness of suitable affine connections. In the final section we discuss the simultaneous integrability of complementary distributions making up an almost product structure.

We introduce Bott connections in general, and we apply them to Lagrangian foliations in particular. This leads to a proof of Weinstein's characterisation of affinely flat manifolds as leaves of Lagrangian foliations. We also prove a Darboux theorem for pairs consisting of a symplectic structure together with a Lagrangian foliation.

4.1 Foliations and Flat Bundles

There are many equivalent ways to define foliations. One of them is the following.

Definition 4.1 A k-dimensional *foliation* $\mathcal{F}$ of an n-manifold M is a partition of M into connected injectively immersed k-manifolds, called the *leaves* of $\mathcal{F}$, which is locally trivial in the following sense: around every point in M there is a chart (U, φ) such that φ sends the connected components of the intersections of the leaves of $\mathcal{F}$ with U to subsets of the form $\mathbb{R}^k \times \{c\}$ in $\mathbb{R}^k \times \mathbb{R}^{n-k} = \mathbb{R}^n$.

The difference $n - k$ is called the codimension of $\mathcal{F}$. The special charts appearing in the definition are called foliation charts.

Another way to phrase the definition is through the existence of an atlas whose transition maps send $\mathbb{R}^k \times \{c\} \subset \mathbb{R}^k \times \mathbb{R}^{n-k}$ to some $\mathbb{R}^k \times \{d(c)\}$ for every c. Note that within a foliation chart there is always a complementary foliation with leaves corresponding to $\{e\} \times \mathbb{R}^{n-k} \subset \mathbb{R}^k \times \mathbb{R}^{n-k}$. However, this is only defined locally. The transition maps of an atlas of foliation charts for $\mathcal{F}$ preserve slices parallel to the first factor in $\mathbb{R}^k \times \mathbb{R}^{n-k}$, but not usually the slices parallel to the second factor.

We have avoided the use of the word submanifold in the definition of a foliation because in general the leaves are not embedded; they are only injectively immersed. Each leaf has an intrinsic topology and manifold structure that is induced by the restrictions of foliation charts, but this intrinsic topology is not the subspace topology of the leaf considered as a subset of M.

Example 4.2   A flow of irrational slope on the square two-torus defines a one-dimensional foliation, in which all leaves are densely injectively immersed – but not embedded – copies of $\mathbb{R}$.

The case of surfaces is very special, because the only non-trivial foliations are those of dimension 1. On a surface every one-dimensional foliation has a complementary one-dimensional foliation, and every foliation is Lagrangian with respect to the symplectic structure given by an area form. Nevertheless, the dynamics of the foliation can be very complicated, and there can be a mix of closed and non-closed orbits, Reeb components, etc.

For a foliation $\mathcal{F}$ on M we denote by $T\mathcal{F}$ the subbundle of TM consisting of vectors tangent to the leaves. This subbundle has the property that its space of sections is closed under taking commutators of vector fields, since the commutator of vector fields tangent to the leaves is again tangent to the leaves. The Frobenius Theorem gives the converse to this statement: an arbitrary subbundle $E \subset TM$ is integrable, i.e. $E = T\mathcal{F}$ for some foliation $\mathcal{F}$ on M, if and only if the space of sections of E is closed under taking commutators. This condition of being closed under brackets can be interpreted as the vanishing of curvature for E. The intrinsic curvature of E is the map

$$c \colon \Lambda^2(E) \longrightarrow TM/E$$
$$X \wedge Y \longmapsto [X, Y] \, .$$

This map vanishes identically if and only if the sections of E are closed under brackets.

The following example provides another connection between flatness and integrability.

Example 4.3   Let $\pi \colon V \longrightarrow B$ be a vector bundle equipped with a connection ∇. Then ∇ defines a vector subbundle $H \subset TV$ that is complementary to the tangent bundle along the fibres of π, denoted $T\pi$. We call $T\pi$, respectively H, the vertical, respectively the horizontal, subbundle of TV. The vertical subbundle is always integrable, and the corresponding foliation of V is the foliation by the fibres. It is a standard theorem in differential geometry that the horizontal subbundle H is integrable if and only if ∇ is flat. In this case the pair (V, ∇) is called a flat (or foliated) bundle. The foliation tangent to H is called the horizontal foliation of the flat bundle. Its leaves are covering spaces of B.

If ∇ is flat, then choosing a base-point $p \in B$ and lifting curves from B to horizontal curves in V, the parallel transport with respect to ∇ depends only on the homotopy class of the curve and so defines a representation

$$hol_\nabla \colon \pi_1(B, p) \longrightarrow GL_k(\mathbb{R}) \,,$$

where k is the rank of V. This is called the holonomy representation of the flat connection ∇.

Conversely, given any representation

$$\rho \colon \pi_1(B, p) \longrightarrow GL_k(\mathbb{R})$$

one can define $V = (\widetilde{B} \times \mathbb{R}^k)/\pi_1(B, p)$, the quotient by the diagonal action, where the fundamental group acts on the universal covering $\widetilde{B}$ by deck transformations and acts on $\mathbb{R}^k$ via ρ. Then V is a rank k vector bundle with a horizontal foliation whose leaves are the images of $\widetilde{B} \times \{x\}$ in V.

In this way one obtains a bijective correspondence between conjugacy classes of (holonomy) representations of $\pi_1(B, p)$ and suitable isomorphism classes of flat bundles.

Example 4.4   In the previous example one can replace the vector bundle V with structure group $GL_k(\mathbb{R})$ by an arbitrary smooth fibre bundle over B with fibre F and structure group the diffeomorphism group $\mathrm{Diff}(F)$. Then isomorphism classes of foliated F-bundles over B correspond to conjugacy classes of representations

$$\rho \colon \pi_1(B, p) \longrightarrow \mathrm{Diff}(F) \,,$$

although there is no linear connection defining the horizontal foliation.

Now let L be a leaf of a k-dimensional foliation $\mathcal{F}$ on M^n. If there exists a tubular neighbourhood of L that is saturated, meaning that it is a union of leaves of $\mathcal{F}$, then the diffeomorphism between the tubular neighbourhood and the total space of the normal bundle of L in M equips the normal bundle with the structure of a foliated bundle. However, in general this is a foliated bundle with structure group $\mathrm{Diff}(\mathbb{R}^{n-k})$, rather than $GL_{n-k}(\mathbb{R})$. In the special situation of Example 4.3, where a horizontal foliation on a vector bundle V is defined by a flat linear connection on $\pi \colon V \longrightarrow B$, the normal bundle of this foliation can be identified with the pullback bundle $\pi^*(V)$, which carries the flat pullback connection.

This idea that the normal bundle to the leaf of a foliation should be foliated, can be abstracted into the notion of a so-called Bott connection on the normal bundle of a foliation. For this, one linearises the discussion to obtain a linear connection defined by the infinitesimal holonomy of an arbitrary foliation. This will be defined throughout, without any assumptions about the existence of saturated tubular neighbourhoods.

For technical reasons we will give a formulation which, instead of the normal bundle, uses the annihilator of $T\mathcal{F}$. This will be useful when we consider Lagrangian foliations on symplectic manifolds.

Definition 4.5 For a foliation $\mathcal{F}$ on M define the *annihilator* as

$$A = \{\, \alpha \in T^*M \mid \alpha|_{T\mathcal{F}} = 0 \,\}.$$

This is a vector subbundle of T^*M whose rank equals the codimension of $\mathcal{F}$. The *Bott connection* on A defined by $\mathcal{F}$ is the map

$$\nabla \colon \Gamma(T\mathcal{F}) \times \Gamma(A) \longrightarrow \Gamma(A)$$
$$(X, \alpha) \longmapsto L_X \alpha.$$

The defining term $L_X \alpha$ equals $i_X d\alpha$ by the Cartan formula, since by definition α annihilates all $X \in T\mathcal{F}$. The assumption that $T\mathcal{F}$ is integrable, i.e. is closed under brackets, ensures that for all $Y \in T\mathcal{F}$ we have

$$(i_X d\alpha)(Y) = -\alpha([X, Y]) = 0,$$

so that $L_X \alpha = i_X d\alpha$ indeed takes values only in the annihilator A of $T\mathcal{F}$.

The Bott connection is clearly bilinear over $\mathbb{R}$, and it satisfies

$$\nabla_X(f\alpha) = L_X(f) \cdot \alpha + f \cdot L_X \alpha = L_X(f) \cdot \alpha + f \cdot \nabla_X \alpha$$

for all smooth functions f. This analogue of the Leibniz rule justifies calling it a connection, although it is not a connection on the vector bundle $A \longrightarrow M$ in the usual sense. The reason is that the covariant derivative ∇_X is defined only if X is tangent to the foliation, so that ∇ is only a partial connection on A. However, it becomes an honest connection whenever A is restricted to a leaf L of $\mathcal{F}$. The name Bott connection is used both for the partial connection defined above and for its restrictions to the leaves.

The following lemma shows that the Bott connection captures the intuitive flatness of the normal bundle.

Lemma 4.6 *For every leaf L of $\mathcal{F}$ the restriction of the Bott connection to $A \longrightarrow L$ is flat.*

Proof Let $X, Y \in \mathfrak{X}(L) = \Gamma(T\mathcal{F}|_L)$. The curvature of ∇ evaluated on X and Y

acts on sections of A as follows:

$$F^\nabla(X, Y)\alpha = \nabla_X\nabla_Y\alpha - \nabla_Y\nabla_X\alpha - \nabla_{[X,Y]}\alpha$$
$$= L_XL_Y\alpha - L_YL_X\alpha - L_{[X,Y]}\alpha = 0$$

by the definition of ∇ and the definition of the commutator $[X, Y]$. □

4.2 Lagrangian Foliations

We now consider foliations on a symplectic manifold (M, ω).

Definition 4.7 A foliation $\mathcal{F}$ on M is called *Lagrangian*, if $T\mathcal{F}$ is a Lagrangian subbundle of the symplectic vector bundle (TM, ω).

If one thinks of the leaves of a foliation as submanifolds (which they are not, in the strictest sense), then a Lagrangian foliation is a foliation by Lagrangian submanifolds. In any case, if $\dim(M) = 2n$, the leaves of a Lagrangian foliation are of dimension n.

Recall from Lemma 2.2 that the symplectic form ω determines an isomorphism $\phi_\omega: TM \longrightarrow T^*M$ via the contraction $\phi_\omega(X) = i_X\omega$.

Lemma 4.8 *For a Lagrangian foliation $\mathcal{F}$, the contraction ϕ_ω restricts to an isomorphism between $T\mathcal{F}$ and the annihilator A.*

Proof Since $T\mathcal{F}$ is Lagrangian, $\phi_\omega(X)$ vanishes on $T\mathcal{F}$ for all $X \in T\mathcal{F}$, and so $\phi_\omega(T\mathcal{F}) \subset A$. However, $T\mathcal{F}$ and A have the same rank n, and ϕ_ω is injective, so ϕ_ω maps $T\mathcal{F}$ isomorphically onto A. □

Under this isomorphism, the Bott connection on A corresponds to a partial connection on $T\mathcal{F}$, again defined only along the leaves of $\mathcal{F}$. Explicitly, for $X, Y \in \Gamma(T\mathcal{F})$ it is given by

$$\nabla_X Y = \phi_\omega^{-1}(L_X\phi_\omega(Y)) = \phi_\omega^{-1}(L_X i_Y\omega) = \phi_\omega^{-1}(i_X di_Y\omega),$$

where the last equality is from the Cartan formula and the assumption that $T\mathcal{F}$ is Lagrangian. Since ω is non-degenerate, this formula is equivalent to the following identity for all $Z \in \mathfrak{X}(M)$:

$$\begin{aligned}\omega(\nabla_X Y, Z) &= (i_X di_Y\omega)(Z) = (di_Y\omega)(X, Z) \\ &= L_X(\omega(Y, Z)) - L_Z(\omega(Y, X)) - \omega(Y, [X, Z]) \qquad (4.1) \\ &= L_X(\omega(Y, Z)) + \omega([X, Z], Y),\end{aligned}$$

where we have used the fact that $\omega(Y, X)$ vanishes because X, Y are in the same Lagrangian subbundle for ω.

We shall also use the name Bott connection for this partial connection on $T\mathcal{F}$. Formula (4.1) is the analogue for the Bott connection of the Koszul formula for the Levi–Civita connection of a Riemannian manifold.

The importance of the Bott connection for the study of Lagrangian foliations stems from the following result.

Proposition 4.9 *For a Lagrangian foliation $\mathcal{F}$ on a symplectic manifold, the Bott connection on $T\mathcal{F}$ is flat and torsion free on every leaf of $\mathcal{F}$.*

Proof The Bott connection on A is flat by Lemma 4.6, and therefore it is flat on $T\mathcal{F}$ as well. It is an instructive exercise to prove flatness directly from (4.1) using the Jacobi identity for the commutator of vector fields.

If T^{∇} denotes the torsion tensor of ∇, then in order to prove that T^{∇} vanishes, we prove that $\omega(T^{\nabla}(X, Y), Z)$ vanishes for all X, Y tangent to $\mathcal{F}$ and arbitrary $Z \in \mathfrak{X}(M)$. First we calculate using (4.1):

$$
\begin{aligned}
\omega(T^{\nabla}(X, Y), Z) &= \omega(\nabla_X Y, Z) \quad - \omega(\nabla_Y X, Z) - \omega([X, Y], Z) \\
&= L_X(\omega(Y, Z)) + \omega([X, Z], Y) - L_Y(\omega(X, Z)) \\
&\quad - \omega([Y, Z], X) - \omega([X, Y], Z) \,.
\end{aligned}
$$

This matches the usual formula for the exterior derivative of ω:

$$
\begin{aligned}
d\omega(X, Y, Z) &= L_X(\omega(Y, Z)) - L_Y(\omega(X, Z)) + L_Z(\omega(X, Y)) \\
&\quad - \omega([X, Y], Z) + \omega([X, Z], Y) - \omega([Y, Z], X) \,.
\end{aligned}
\tag{4.2}
$$

The two sums are the same, except for the term $L_Z(\omega(X, Y))$, which vanishes because X and Y are in the same Lagrangian subbundle with respect to ω. We conclude

$$
\omega(T^{\nabla}(X, Y), Z) = d\omega(X, Y, Z) \,,
$$

and this vanishes because ω is closed. Thus the Bott connection ∇ on $T\mathcal{F}$ is torsion free. □

Corollary 4.10 *If L is a leaf of a Lagrangian foliation, then $TL \longrightarrow L$ admits a torsion-free flat connection. In other words, L is an affinely flat manifold.*

This shows that the leaves of Lagrangian foliations are very special.

Example 4.11 A closed oriented surface is affinely flat if and only if it is a torus. The torus is flat even in the metric, Riemannian, sense. It is a theorem of Milnor and Benzecri that surfaces of genus different from 1 do not admit flat affine connections. This is not a trivial consequence of the Gauss–Bonnet Theorem, since the connections in question are not assumed to be metric.

Recall that if we consider just Lagrangian submanifolds, without requiring that they be leaves of Lagrangian foliations, then there are no restrictions on the manifolds.

Example 4.12   We saw in Subsection 3.1.2 of Chapter 3 that the cotangent bundle of any smooth manifold is symplectic, and that the zero-section is a Lagrangian submanifold.

Corollary 4.10 admits the following sharpening and converse, originally due to Weinstein [Wei-71].

Theorem 4.13   *A smooth manifold is a leaf of a Lagrangian foliation if and only if it admits a torsion-free flat affine connection.*

Proof   One direction was already proved in Corollary 4.10. For the converse assume that M admits a torsion-free flat affine connection ∇. We consider the cotangent bundle T^*M with its canonical symplectic form ω_{can} defined in Subsection 3.1.2. Without using ∇, we know that the zero-section is a Lagrangian submanifold. However, we now want to prove more, namely that this Lagrangian submanifold is actually a leaf of a Lagrangian foliation on T^*M.

Instead of working with ∇ on TM, we use the dual connection ∇^* on T^*M. This is flat because ∇ is. Flatness means that there exist local trivialisations by parallel sections. In the case of ∇^* on T^*M we can trivialise T^*M by ∇^*-parallel one-forms $\alpha_1, \ldots, \alpha_n$. The horizontal subbundle for ∇^* is integrable, and locally the leaves of the corresponding foliations are just the graphs of constant linear combinations of the parallel one-forms α_i. Now because ∇ is torsion free, every ∇^*-parallel one-form is closed because for a ∇-parallel one-form α we have $d\alpha(X, Y) = \alpha(T^\nabla(X, Y))$. Thus constant linear combinations of the α_i are also closed. Therefore, locally, the leaves of the horizontal foliation are graphs of closed one-forms and are therefore Lagrangian by Proposition 3.20.   □

To end this section we want to prove the following Darboux Theorem for Lagrangian foliations.

Proposition 4.14   *Let (M, ω) be a symplectic manifold with a Lagrangian foliation $\mathcal{F}$. Then locally, around an arbitrary point $x \in M$, there are coordinates $(q_1, \ldots, q_n, p_1, \ldots, p_n)$ such that $\omega = \sum_i dq_i \wedge dp_i$ and, at the same time, the leaves of $\mathcal{F}$ are given by setting $p_1, \ldots, p_n$ equal to constants.*

In other words, any Lagrangian foliation is locally standard, in that there is a foliation-preserving symplectomorphism with a suitable open set in a cotangent bundle foliated by its fibres. We need the following lemma, which one can think of as an embellished Poincaré Lemma.

Lemma 4.15 *Let (M, ω) be a 2n-dimensional symplectic manifold and $\mathcal{F}$ a Lagrangian foliation on M. Around every point $x \in M$ there exist an open neighbourhood U, diffeomorphic to $\mathbb{R}^{2n}$, and a one-form σ on U, such that $\omega|_U = d\sigma$ and $\sigma|_{T\mathcal{F}} = 0$, i.e. $\sigma(Y) = 0$ for every tangent vector $Y \in T\mathcal{F}$.*

Proof Let U be an open neighbourhood of x, diffeomorphic to $\mathbb{R}^{2n} = \mathbb{R}^n \times \mathbb{R}^n$, forming a foliation chart for $\mathcal{F}$ so that the leaves of $\mathcal{F}$ in U are given by $\mathbb{R}^n \times \{c\}$. Fix a Lagrangian submanifold $L \cong \mathbb{R}^n \subset U$ through x, intersecting each leaf of $\mathcal{F}$ restricted to U once and transversely. Choosing coordinates on U suitably we can assume that $L = \{0\} \times \mathbb{R}^n$.

Define the homotopy

$$\phi \colon U \times [0, 1] \longrightarrow U$$

$$(q, p, t) \longmapsto \phi_t(q, p) = ((1 - t)q, p),$$

where $q, p \in \mathbb{R}^n$. Then $\phi_t \colon U \longrightarrow U$ is a diffeomorphism for all $t < 1$, $\phi_0 = \mathrm{Id}_U$ and ϕ_1 has image in L. The homotopy ϕ preserves the foliation $\mathcal{F}$ on U. For $t \in [0, 1)$ we can define a time-dependent vector field X_t on U, tangent to the foliation $\mathcal{F}$, by

$$X_t(y) = \left.\frac{d}{ds}\right|_{s=t} \phi_s\left(\phi_t^{-1}(y)\right) \quad \forall y \in U.$$

Then

$$\left.\frac{d}{ds}\right|_{s=t} \phi_s = X_t \circ \phi_t,$$

i.e. the family of diffeomorphisms ϕ_t for $t \in [0, 1)$ is the flow of X_t. Notice that X_1 is not defined.

For $t \in [0, 1]$ define a one-form σ_t on U by

$$\sigma_t(Y_z) = -\omega_{\phi_t(z)}\left(\left.\frac{d}{ds}\right|_{s=t} \phi_s(z), (\phi_t)_*(Y_z)\right)$$

for all $z \in U$ and tangent vectors $Y_z \in T_z M$. For $t < 1$ we can write

$$\sigma_t = -\phi_t^*(i_{X_t}\omega).$$

Since the vector

$$\left.\frac{d}{ds}\right|_{s=t} \phi_s(z)$$

is tangent to the Lagrangian foliation $\mathcal{F}$ for all $z \in U$ and $t \in [0, 1]$, it follows that $\sigma_t|_{T\mathcal{F}} = 0$ for all $t \in [0, 1]$.

For $t \in [0, 1)$ the Cartan formula shows that

$$\left.\frac{d}{ds}\right|_{s=t} \phi_s^*\omega = \phi_t^*(L_{X_t}\omega) = d\phi_t^*(i_{X_t}\omega) = -d\sigma_t.$$

Since σ_t is a smooth family of one-forms on U, this equation also holds for $t = 1$. Integration then yields

$$\phi_0^*\omega - \phi_1^*\omega = d\sigma,$$

where

$$\sigma = \int_0^1 \sigma_t \, dt$$

is a one-form on U. However, $\phi_1^*\omega$ vanishes since L is Lagrangian, and $\phi_0 = \mathrm{Id}_U$. Thus we conclude $\omega = d\sigma$, and by construction $\sigma|_{T\mathcal{F}} = 0$.   □

Proof of Proposition 4.14   We use the Moser method as in the proof of Theorem 3.15. Let U be an open neighbourhood of x, diffeomorphic to $\mathbb{R}^{2n} = \mathbb{R}^n \times \mathbb{R}^n$, forming a foliation chart for $\mathcal{F}$, so that the leaves of $\mathcal{F}$ restricted to U are given by $\mathbb{R}^n \times \{c\}$. Let $q_1, \dots, q_n$ be coordinates on the first $\mathbb{R}^n$ factor and $p_1, \dots, p_n$ on the second factor. Changing coordinates by a linear isomorphism of $\mathbb{R}^{2n}$ we can assume that at the point x

$$\omega(x) = \sum_i dq_i \wedge dp_i \, .$$

On U consider the symplectic forms $\omega_1 = \omega|_U$ and

$$\omega_0 = \sum_i dq_i \wedge dp_i \, .$$

According to Lemma 4.15 we can find a one-form σ_1 on U such that $\omega_1 = d\sigma_1$ and σ_1 vanishes on the leaves of $\mathcal{F}$. Moreover, $\omega_0 = d\sigma_0$ with $\sigma_0 = \sum_i q_i dp_i$, hence σ_0 also vanishes on the leaves of $\mathcal{F}$. Set $\sigma = \sigma_1 - \sigma_0$. Then

$$\omega_1 - \omega_0 = d\sigma, \quad \sigma|_{T\mathcal{F}} = 0 \, .$$

For $t \in [0, 1]$ let

$$\omega_t = \omega_0 + t(\omega_1 - \omega_0) = \omega_0 + td\sigma \, .$$

Since at the point $x \in M$ we have $\omega_1(x) = \omega_0(x)$, it follows that $\omega_t(x) = \omega_0(x)$ is non-degenerate for all $t \in [0, 1]$. Shrinking U if necessary, we can assume that ω_t is symplectic on U for all $t \in [0, 1]$. The foliation $\mathcal{F}$ is Lagrangian for ω_t for all $t \in [0, 1]$.

Define a time-dependent vector field X_t on U by

$$i_{X_t}\omega_t = -\sigma \, .$$

This X_t is tangent to the foliation $\mathcal{F}$ for all $t \in [0, 1]$ because $\sigma|_{T\mathcal{F}}$ vanishes and $T\mathcal{F}$ is maximal isotropic. Let ψ_t denote the flow of X_t. As in the proof of Theorem 3.15 it follows that $\psi_1^*\omega_1 = \omega_0$. Since X_t is tangent to $\mathcal{F}$ for all times

$t \in [0, 1]$, the diffeomorphism ψ_1 preserves the foliation $\mathcal{F}$. This completes the proof. ☐

4.3 Integrability and Torsion

We have already related the integrability of subbundles in the tangent bundle to the absence of curvature. One can also relate integrability to the absence of torsion, and this is what we do now.

The basic observation in this direction is the following. As usual, we say that ∇ preserves a subbundle $E \subset TM$ if $\nabla_X Y \in \Gamma(E)$ for all $Y \in \Gamma(E)$ and arbitrary X. In this case one also says that E is ∇-parallel.

Lemma 4.16 *If a subbundle $E \subset TM$ is preserved by a torsion-free affine connection on M, then E is integrable.*

Proof Suppose X and Y are sections of E. The torsion-freeness of ∇ means

$$[X, Y] = \nabla_X Y - \nabla_Y X .$$

The two summands on the right-hand side are in E because we assumed that E is preserved by ∇. Thus we conclude that $\Gamma(E)$ is closed under taking commutators. ☐

We can elaborate on this to obtain a characterisation of integrability through the existence of certain torsion-free connections.

Proposition 4.17 *For a subbundle $E \subset TM$ the following are equivalent:*

 (i) *the subbundle $E \subset TM$ is integrable,*
 (ii) *the bundle $E \longrightarrow M$ has a torsion free connection,*
(iii) *the manifold M admits a torsion free affine connection preserving E.*

Proof The third condition implies the second one by restricting the connection, and the second condition implies the first one by the argument in the proof of the lemma above. That proof required a torsion-free connection only on E, not on all of TM.

The proof will be complete once we show that (i) implies (iii). So assume that E is integrable, i.e. $E = T\mathcal{F}$ for some foliation $\mathcal{F}$. We can cover M by foliation charts for $\mathcal{F}$, so that in each of these charts the leaves of $\mathcal{F}$ correspond to the parallel copies of the first factor in the product decomposition $\mathbb{R}^k \times \mathbb{R}^{n-k} = \mathbb{R}^n$. Locally in these charts we define torsion free flat connections by declaring the coordinate vector fields to be parallel. Then these locally defined connections preserve E, since it is just the span of the first k coordinate vector fields, which are by definition parallel. Now patch together these locally defined connections using a partition of unity. The resulting connection is still torsion free and preserves E, since these properties survive the formation of convex combinations. (Of course the resulting connection is not usually flat, since flatness does not survive under convex combinations.) Thus we have proved that (i) implies (iii).   □

We want to adapt this result to the case of Lagrangian foliations on symplectic manifolds. In this case we consider ω-compatible connections defined as follows.

Definition 4.18   Let (M, ω) be a symplectic manifold. An affine connection ∇ on M is *compatible with ω* if

$$L_X(\omega(Y,Z)) = \omega(\nabla_X Y, Z) + \omega(Y, \nabla_X Z) \tag{4.3}$$

for all $X, Y, Z \in \mathfrak{X}(M)$.

Such ω-compatible connections are sometimes called symplectic connections. Using these connections we have the following characterisation of integrability for Lagrangian subbundles.

Proposition 4.19   *Let (M, ω) be a symplectic manifold. For a Lagrangian subbundle $E \subset TM$ the following are equivalent:*

 (i) *the subbundle $E \subset TM$ is integrable,*
 (ii) *the bundle $E \longrightarrow M$ has a torsion-free connection,*
 (iii) *the manifold M admits a torsion-free affine connection compatible with ω that preserves E.*

Proof   The implications from (iii) to (ii) and from (ii) to (i) are as in the proof of Proposition 4.17. For the implication from (i) to (iii) we can also consult that proof, and we see that the novelty is in ensuring that the connection is compatible with ω. For this we can use the Darboux Theorem for Lagrangian foliations proved in Proposition 4.14. By that result we may cover M by Darboux charts for ω that have the additional property that the Lagrangian foliation $\mathcal{F}$ with $T\mathcal{F} = E$ is standard, given by the first factor in $\mathbb{R}^n \times \mathbb{R}^n = \mathbb{R}^{2n}$. Then the torsion-free connection, defined on the domain of such a chart by the requirement that the coordinate vector fields $\frac{\partial}{\partial p_i}$ and $\frac{\partial}{\partial q_i}$ are parallel, preserves E and is compatible with ω. Patching together with a partition of unity we obtain a globally defined affine connection. This is still torsion free, compatible with ω, and preserves the subbundle E.   □

Note that within each of these special Darboux charts there is a Lagrangian foliation complementary to $\mathcal{F}$. However, the transition maps between these charts preserve only $\mathcal{F}$ and usually do not preserve the Lagrangian complements singled out by the Darboux coordinates.

The special connections for Lagrangian foliations that we just constructed have the additional property that they restrict to the Bott connection along the foliation. This may be surprising since the Bott connection is unique, and the connections we have just constructed are certainly not unique.

Proposition 4.20 *Let (M, ω) be a symplectic manifold with a Lagrangian foliation $\mathcal{F}$. Any ω-compatible torsion-free connection ∇ that preserves $T\mathcal{F}$ restricts to $T\mathcal{F}$ as the Bott connection.*

Proof Recall from (4.1) that for $X, Y \in T\mathcal{F}$ the Bott connection ∇^B is given implicitly by

$$\omega(\nabla^B_X Y, Z) = L_X(\omega(Y, Z)) + \omega([X, Z], Y) \,.$$

We can compare this with the ω-compatibility of ∇, which by (4.3) is equivalent to

$$\omega(\nabla_X Y, Z) = L_X(\omega(Y, Z)) + \omega(\nabla_X Z, Y) \,.$$

The torsion freeness of ∇ gives

$$\omega(\nabla_X Z, Y) = \omega(\nabla_Z X, Y) + \omega([X, Z], Y) \,,$$

where the first term on the right-hand side vanishes because X and Y are tangent to $\mathcal{F}$, ∇ preserves $T\mathcal{F}$, and $T\mathcal{F}$ is Lagrangian. Thus we conclude that the implicit formulas above for $\nabla^B_X Y$ and for $\nabla_X Y$ agree. $\qquad\qquad\square$

Finally we would like to work out the special properties of differences of torsion-free symplectic connections preserving a given Lagrangian foliation $\mathcal{F}$. If ∇ and ∇' are two affine connections on M, then their difference $A = \nabla' - \nabla$ is a one-form with values in the endomorphisms of the tangent bundle. We shall write this as

$$A(X, Y) = \nabla'_X Y - \nabla_X Y \,.$$

The endomorphism-valued one-form A determines a trilinear map

$$\omega_A : \mathfrak{X}(M) \times \mathfrak{X}(M) \times \mathfrak{X}(M) \longrightarrow C^\infty(M)$$
$$(X, Y, Z) \longmapsto \omega(A(X, Y), Z) \,.$$

Conversely, because ω is non-degenerate, ω_A determines A.

Using this notation we prove the following characterisation of symplectic torsion-free connections preserving $\mathcal{F}$.

Theorem 4.21 *Let (M, ω) be a symplectic manifold with a Lagrangian foliation $\mathcal{F}$. There exist ω-compatible torsion-free connections ∇ which preserve $T\mathcal{F}$. The set of all such connections is naturally an affine space whose vector space of translations is the space of endomorphism-valued one-forms $A \in \Omega^1(\mathrm{End}(TM))$ that have the property that*

(i) ω_A *is symmetric under permutations of all three arguments,*

(ii) ω_A *vanishes on triples for which at least two of the vectors are in $T\mathcal{F}$.*

Proof The existence of such connections was established in the proof of Proposition 4.19.

Let A be the difference of two such connections. Because the two connections have the same torsion, we conclude that A is symmetric, equivalently ω_A is symmetric in its first two arguments. Writing out the ω-compatibility of each of the two connections as in (4.3) and taking the difference of the two identities, we see that $\omega(A(X, Y), Z) = \omega(A(X, Z), Y)$; in other words ω_A is symmetric in the second and third arguments. Thus, combining the two symmetries, ω_A satisfies (i).

Since both connections preserve $T\mathcal{F}$, we know that $A(X, Y)$ is in $T\mathcal{F}$ as soon as Y is. As we have already proved that (i) holds, if two of the three arguments are in $T\mathcal{F}$, we may assume that they are Y and Z. Then $A(X, Y)$ and Z are both in $T\mathcal{F}$ and $\omega_A(X, Y, Z) = 0$ because $T\mathcal{F}$ is Lagrangian with respect to ω. This proves (ii).

We have shown that the difference of two torsion-free ω-compatible connections preserving $T\mathcal{F}$ satisfies (i) and (ii). These conditions define a linear subspace in $\Omega^1(\mathrm{End}(TM))$. Conversely, if we modify a given connection with the desired properties by the addition of an element from this subspace, then one can check directly that all the desired properties survive. □

4.4 Almost Product Structures

Instead of the integrability of a single subbundle $E \subset TM$ we now want to discuss the simultaneous integrability of two complementary subbundles. So we consider a fixed splitting $TM = F \oplus G$ of the tangent bundle of a smooth manifold M into complementary smooth subbundles F and G.

This situation is often encoded using the notion of an almost product structure, which is obtained from that of an almost complex structure by a sign change.

Definition 4.22 An *almost product structure* on M is a smooth section $I \in \Gamma(\text{End}(TM))$ with the property that $I^2 = \text{Id}_{TM}$ but $I \neq \pm\,\text{Id}_{TM}$.

We make the requirement $I \neq \pm\,\text{Id}_{TM}$ in order to avoid the consideration of trivial special cases.

Lemma 4.23 *An almost product structure I defines a unique splitting $TM = F \oplus G$ into complementary smooth subbundles. Conversely, every such splitting arises from an almost product structure I, which is unique up to sign.*

Proof Given an almost product structure I, we can consider its ± 1-eigenspaces in every tangent space. These are complementary, and their dimensions are both semicontinuous in the sense that each can increase only on a closed subset. It follows that on a connected manifold M the dimensions are constant, and therefore the eigenspaces are the fibres of constant-rank smooth subbundles of TM.

Conversely, given a smooth splitting $TM = F \oplus G$, we can define $I = \pm(\text{Id}_F \ominus \text{Id}_G)$. This is a smooth section of the endomorphism bundle well defined up to sign, which is obviously an almost product structure. □

Definition 4.24 We will say that an almost product structure I is *integrable* if its eigenbundles F and G are both integrable to foliations $\mathcal{F}$ and $\mathcal{G}$. The two foliations are then everywhere complementary and form a *bifoliation*.

The following is the extension of Proposition 4.17 to this situation.

Proposition 4.25 *For an almost product structure $I \in \Gamma(\text{End}(TM))$ the following are equivalent:*

(i) *I is integrable to a bifoliation,*
(ii) *each of the eigenbundles of I has a torsion-free connection,*
(iii) *the manifold M admits a torsion-free affine connection preserving the splitting of TM into the eigenbundles of I,*
(iv) *the manifold M admits a torsion-free affine connection ∇ whose covariant derivative commutes with I, i.e. $\nabla_X(IY) = I(\nabla_X Y)$ for all tangent vectors X and Y.*

Proof The equivalence of (i) and (ii) is immediate from Proposition 4.17.

If (ii) holds, then one could try to prove (iii) by taking the direct sum connection of the two connections on the subbundles. However, such a direct sum connection will not be torsion free, unless we are careful with the choice of connections on the subbundles. We therefore proceed as follows.

Let F and G be the eigenbundles for I. If F has a torsion-free connection ∇, we first modify it to a new connection ∇^F such that $\nabla^F_X Y$ agrees with $\nabla_X Y$ if

X is in F, and equals $[X, Y]_F$, the F-component of the commutator, if X is in G. This defines a connection, which is torsion free, because the terms relevant for the torsion have not been modified, and ∇ was assumed to be torsion free. Similarly we modify any torsion-free connection on G to a new one, called ∇^G, which satisfies $\nabla^G_X Y = [X, Y]_G$ if X is in F and Y is in G. Finally, consider the direct sum connection $\nabla^F \oplus \nabla^G$. By the definition, this preserves F and G. Moreover, if both X and Y are in F, or both are in G, then the torsion $T(X, Y)$ vanishes because both ∇^F and ∇^G are torsion free. This leaves us to check the mixed torsion $T(X, Y)$ for $X \in \Gamma(F)$ and $Y \in \Gamma(G)$. In this case we have:

$$
\begin{aligned}
T(X, Y) &= (\nabla^F \oplus \nabla^G)_X Y - (\nabla^F \oplus \nabla^G)_Y X - [X, Y] \\
&= \nabla^G_X Y - \nabla^F_Y X - [X, Y] \\
&= [X, Y]_G - [Y, X]_F - [X, Y] \\
&= [X, Y]_G + [X, Y]_F - [X, Y] \\
&= 0 \,.
\end{aligned}
$$

This completes the proof that (ii) implies (iii).

Conversely, if (iii) holds, then the given torsion-free connection restricts to torsion-free connections on the eigenbundles of I. Finally, the condition that ∇ commutes with I is just another way of saying that ∇ preserves the eigenbundles of I, and so (iv) is equivalent to (iii). □

Notes for Chapter 4

1. Corollary 4.10 and Theorem 4.13 are due to Weinstein [Wei-71].
2. Different proofs of Proposition 4.14 can be found in [Wei-71, Vai-89].
3. Our treatment of Bott connections for Lagrangian foliations is similar to that of Forger–Yepes [FY-13].

5

Künneth Structures

With this chapter we begin the main part of our discussion of Künneth geometry. We first define Künneth and almost Künneth structures. The former are the main structures whose geometry we investigate in this and the coming chapters. In this chapter we give only their most basic properties, and we discuss their automorphism groups. Most of this chapter consists of the discussion of several classes of examples.

5.1 Basic Definitions and Results

If a smooth manifold M has a product structure $M = X \times Y$, then the differential forms on it have a bigrading defined by the decomposition

$$\Lambda^k(M) = \bigoplus_{p+q=k} \Lambda^p(X) \otimes \Lambda^q(Y) . \tag{5.1}$$

The exterior derivative d splits as $d_X + d_Y$ with respect to the product structure.

This structure is still present if M is only locally a product as above, which means that M has two complementary smooth foliations $\mathcal{F}$ and $\mathcal{G}$. Again the exterior derivative d splits as $d_{\mathcal{F}} + d_{\mathcal{G}}$ in the obvious way. Note that the type decomposition applied to $d^2 = 0$ gives the following:

$$d_{\mathcal{F}}^2 = 0 , \qquad d_{\mathcal{F}} d_{\mathcal{G}} + d_{\mathcal{G}} d_{\mathcal{F}} = 0 , \qquad d_{\mathcal{G}}^2 = 0 . \tag{5.2}$$

Here is the central definition, which we learned from Guillemin [Gui-00].

Definition 5.1 A *Künneth structure* on a smooth manifold M consists of two complementary smooth foliations $\mathcal{F}$ and $\mathcal{G}$ and a symplectic form ω of type $(1, 1)$.

The name comes from the analogy with Kähler manifolds, which also have

57

a symplectic form of type $(1, 1)$ (for a different bigrading) and the association of the name Künneth with the discussion of Cartesian products in algebraic topology. Here the Cartesian product is present only locally, as the local product structure provided by the bifoliation formed by $\mathcal{F}$ and $\mathcal{G}$.

Instead of focusing on the type decomposition of the symplectic form, it is often convenient to work with the properties of the foliations, using the following reformulation of the definition.

Lemma 5.2 *A Künneth structure is equivalent to a symplectic structure together with a pair of complementary Lagrangian foliations.*

Proof In one direction, if $(\mathcal{F}, \mathcal{G}, \omega)$ is Künneth, then the symplectic form ω being of type $(1, 1)$ forces the two foliations to be isotropic. As ω is non-degenerate, the foliations are equidimensional and therefore Lagrangian.

Conversely, suppose (M, ω) is symplectic and admits a pair of complementary Lagrangian foliations $\mathcal{F}$ and $\mathcal{G}$. Then it follows that ω is of type $(1, 1)$. $\quad\square$

This explains why Künneth structures are often called bi-Lagrangian structures. We will use both names.

We saw in Proposition 4.14 that there is a Darboux theorem for symplectic forms with a single Lagrangian foliation, showing that both the symplectic form and a Lagrangian foliation can be made to look locally standard in suitable charts. This is no longer true for pairs of Lagrangian foliations. Given a Künneth structure $(\mathcal{F}, \mathcal{G}, \omega)$ we can of course apply Proposition 4.14 to ω and $\mathcal{F}$, say, but we will have no control over $\mathcal{G}$ in those special charts. Alternatively, we can use bifoliation charts that make $\mathcal{F}$ and $\mathcal{G}$ simultaneously standard, so that their leaves look like the parallel copies of the two factors in $\mathbb{R}^n \times \mathbb{R}^n$. In such a chart the symplectic form can be arranged to have the following nice form.

Proposition 5.3 *Let M be a manifold with a Künneth structure $(\mathcal{F}, \mathcal{G}, \omega)$. Then every point $p \in M$ has an open neighbourhood U with a smooth function $\rho \colon U \to \mathbb{R}$ such that $\omega|_U = d_{\mathcal{F}} d_{\mathcal{G}}(\rho)$.*

Proof We work locally on a contractible open neighbourhood of p. Since ω is closed we may choose a local primitive α by the Poincaré Lemma. We decompose α into its $(1, 0)$ and its $(0, 1)$ parts:

$$\alpha = \alpha^{1,0} + \alpha^{0,1} \, .$$

Applying $d = d_{\mathcal{F}} + d_{\mathcal{G}}$ we find

$$\omega = d\alpha = d_{\mathcal{F}} \alpha^{1,0} + (d_{\mathcal{F}} \alpha^{0,1} + d_{\mathcal{G}} \alpha^{1,0}) + d_{\mathcal{G}} \alpha^{0,1} \, .$$

Since ω is assumed to be of type $(1, 1)$, the first and third terms on the right-hand side must vanish, as they are of type $(2, 0)$, and of type $(0, 2)$ respectively. By the Poincaré Lemma with parameters applied along the leaves of $\mathcal{F}$ we find that there is a function f such that $d_{\mathcal{F}} f = \alpha^{1,0}$. Similarly, the Poincaré Lemma applied along the leaves of $\mathcal{G}$, with parameters along $\mathcal{F}$, gives a function g such that $d_{\mathcal{G}} g = \alpha^{0,1}$. Thus we obtain

$$\omega = d_{\mathcal{F}} \alpha^{0,1} + d_{\mathcal{G}} \alpha^{1,0} = d_{\mathcal{F}} d_{\mathcal{G}} g + d_{\mathcal{G}} d_{\mathcal{F}} f = d_{\mathcal{F}} d_{\mathcal{G}} (g - f)$$

since $d_{\mathcal{F}}$ and $d_{\mathcal{G}}$ anti-commute by (5.2). □

The function ρ plays the rôle of the Kähler potential in Kähler geometry, and provides local invariants of the Künneth structure. We will see in Chapter 7 that these can be interpreted through the curvature of a naturally defined connection.

For now let us observe that ρ can always be modified by adding functions that depend only on the variables along $\mathcal{F}$, and functions that depend only on the variables along $\mathcal{G}$. Furthermore, one can simplify ρ by applying local diffeomorphisms preserving the two foliations. If we write $x = (x_1, \ldots, x_n)$ for local coordinates along $\mathcal{F}$ and $y = (y_1, \ldots, y_n)$ for local coordinates along $\mathcal{G}$, then by the Linear Darboux Theorem we may assume that

$$\rho(x, y) = \sum_{i=1}^{n} x_i y_i + \text{higher order terms} .$$

The condition that ω is standard, i.e. $\omega = \sum_{i=1}^{n} dx_i \wedge dy_i$, in this bifoliation chart for $\mathcal{F}$ and $\mathcal{G}$ amounts to the vanishing of the higher-order terms in the potential ρ. In this case the Künneth structure is locally isomorphic to the standard structure on $\mathbb{R}^n \times \mathbb{R}^n$. We shall see in Theorem 7.6 that this happens if and only if the naturally defined connection is flat.

Note that a Künneth structure on a manifold M defines on the tangent bundle TM the structure of a Künneth vector bundle in the sense of Definition 2.35. This situation is important enough to deserve a name of its own.

Definition 5.4 An *almost Künneth structure* on a smooth manifold M consists of a non-degenerate two-form ω together with two complementary ω-isotropic subbundles F and G of TM.

Since ω is assumed non-degenerate, the dimension of M is even, say equal to $2n$. Moreover, the rank of any isotropic subbundle of TM is at most $\frac{1}{2} \dim(M) = n$. Therefore, the existence of two complementary isotropic subbundles implies that they are both of maximal rank n. We will usually call them Lagrangian, as in the case when ω is symplectic. Recall that a compatible almost complex structure can be chosen to give an isomorphism between F and G, so the two Lagrangian subbundles are always isomorphic to each other.

Weakening the notion of a Künneth structure to an almost Künneth structure we drop both the requirement that ω be closed, and the assumption that the Lagrangian subbundles are integrable. If these requirements are satisfied, so that the almost Künneth structure is in fact induced by a Künneth structure, then we say that the almost Künneth structure is integrable. There are geometric situations in which partially integrable almost Künneth structures appear naturally. For example, the two-form might be symplectic, and one of the two Lagrangian distributions may be integrable, but the second one is not.

Proposition 2.36 tells us that the existence of an almost Künneth structure on M is equivalent to the existence of an almost symplectic form admitting a Lagrangian subbundle. By the same proposition, the existence of an almost Künneth structure can be reformulated as follows.

Lemma 5.5 *An almost Künneth structure on a 2n-manifold M is equivalent to an almost complex structure J on M admitting a totally real subbundle of rank n.*

Recall that a subbundle $F \subset TM$ is totally real with respect to J if $J(F) \cap F = 0$.

Example 5.6 Not every symplectic vector bundle admits a Lagrangian subbundle. For example, any area form on S^2 makes the tangent bundle $TS^2 \longrightarrow S^2$ into a symplectic vector bundle. However, by the Hairy Ball Theorem, this cannot have any rank 1 subbundle.

This is a special case of Corollary 2.39, which in the situation when the Künneth vector bundle is the tangent bundle tells us the following.

Theorem 5.7 *If a closed smooth manifold M of dimension 4k + 2 admits an almost Künneth structure, then its Euler characteristic $\chi(M)$ vanishes.*

This is just the first of many strong constraints on the topology of a manifold imposed by the existence of an (almost) Künneth structure. The conclusion is false in dimension $4k$, as we will see in Chapter 10.

We will always consider (almost) Künneth manifolds as oriented manifolds, oriented by the non-degenerate two-form ω. However, it may happen that the Lagrangian subbundle L is non-orientable. Therefore, taking up Definition 2.37, we will use the following terminology.

Definition 5.8 An almost Künneth structure (ω, F, G) on a manifold M is *orientable* if the Lagrangian subbundles F and G are orientable vector bundles.

Orientability is a very minor restriction, since it can be ensured by passing to a double covering if necessary.

5.2 The Induced Metric and the Automorphism Group

For every almost Künneth structure (ω, F, G) we have an underlying almost product structure

$$I: TM \longrightarrow TM$$
$$X_F + X_G \longmapsto X_F - X_G \, .$$

In other words, with respect to the splitting $TM = F \oplus G$, we have $I = \mathrm{Id}_F \ominus \mathrm{Id}_G$. This almost product structure, together with ω, allows us to define a natural pseudo-Riemannian metric associated with the almost Künneth structure.

Proposition 5.9 *For any almost Künneth structure (ω, F, G) the map*

$$g: TM \times TM \longrightarrow \mathbb{R}$$
$$(X, Y) \longmapsto \omega(IX, Y)$$

defines a pseudo-Riemannian metric of neutral signature (n, n).

Proof Since I is linear and ω is bilinear, it is clear that g is bilinear. Moreover, g is non-degenerate because I is an isomorphism and ω is non-degenerate. To check the symmetry of g we use the skew-symmetry of ω and the assumption that F and G are ω-isotropic:

$$
\begin{aligned}
g(X, Y) &= \omega(IX, Y) \\
&= \omega(X_F - X_G, Y_F + Y_G) \\
&= \omega(X_F, Y_G) - \omega(X_G, Y_F) \\
&= -\omega(Y_G, X_F) + \omega(Y_F, X_G) \\
&= \omega(Y_F - Y_G, X_F + X_G) \\
&= \omega(IY, X) \\
&= g(Y, X) \, .
\end{aligned}
$$

We have now proved that g is a pseudo-Riemannian metric. Note that F and G are ω-isotropic and I-invariant, and hence null for g. Therefore the signature of g is (n, n). □

The definitions of Künneth and almost Künneth structures treat the two Lagrangian subbundles of the tangent bundle symmetrically. However, this symmetry is broken when writing down the almost product structure I, for one then has to decide on which summand I is the identity and on which it is minus the identity. Of course both choices work equally well, and therefore the metric g is well defined only up to an overall sign.

It will be very useful to consider the group of structure-preserving diffeomorphisms of an almost Künneth structure.

Definition 5.10 For an almost Künneth structure (ω, F, G) on a smooth manifold M, its (almost Künneth) *automorphism group* is

$$\text{Aut}(\omega, F, G) = \{\varphi \in \text{Diff}(M) \mid \varphi^*\omega = \omega \,,\; \varphi_* F = F \,,\; \varphi_* G = G\}\,.$$

The condition $\varphi^*\omega = \omega$ implies that automorphisms are volume preserving, and so they preserve the orientation of M defined by the volume form ω^n. However, it may happen that an automorphism reverses the orientations of the Lagrangian subbundles F and G – assuming they are orientable to begin with.

As we have noted several times already, the vector bundles F and G are isomorphic to each other. In particular, they are either both orientable, or both non-orientable. If an automorphism preserves a given orientation on one of these subbundles, then because it preserves the orientation of M, it also preserves the orientation of the other subbundle.

Definition 5.11 For an orientable almost Künneth structure (ω, F, G) on a smooth manifold M, its *orientation-preserving automorphism group* is

$$\text{Aut}_+(\omega, F, G) = \{\varphi \in \text{Aut}(\omega, F, G) \mid \varphi_* \text{ is orientation-preserving on } F\}\,.$$

Note that this is a subgroup of index at most 2 in the full automorphism group $\text{Aut}(\omega, F, G)$.

According to Proposition 5.9 every almost Künneth structure defines an essentially unique pseudo-Riemannian metric g, which must be preserved by all almost Künneth automorphisms. This implies the following:

Theorem 5.12 *The automorphism group of any almost Künneth structure is a group of isometries of the associated pseudo-Riemannian metric, and is therefore a finite-dimensional Lie group.*

Since the associated metric is indefinite, its isometry group is usually non-compact, even on compact manifolds. Examples of this will appear in Corollary 5.19 below. Nevertheless, Theorem 5.12 shows that automorphism groups of (almost) Künneth structures, although not compact, are much smaller than more general groups of symplectomorphisms or volume-preserving diffeomorphisms.

5.3 Examples

It is obvious that $\mathbb{R}^{2n}$ has a Künneth structure for which the leaves of the foliations $\mathcal{F}$ and $\mathcal{G}$ are the copies $\mathbb{R}^n \times \{y\}$ and $\{x\} \times \mathbb{R}^n$ respectively, of the two factors in the product decomposition $\mathbb{R}^{2n} = \mathbb{R}^n \times \mathbb{R}^n$, together with the standard symplectic form $\omega = \sum_{i=1}^{n} dx_i \wedge dy_i$. Here the x_i are coordinates on the first factor and the y_j are coordinates on the second factor.

This structure is invariant under translations, and therefore descends to the torus $T^{2n} = \mathbb{R}^{2n}/\mathbb{Z}^{2n}$. This is our first example of a Künneth structure on a closed manifold.

In dimension 2 the only closed symplectic manifold admitting a foliation is the torus. This carries the obvious Künneth structure discussed above, given by its global product decomposition and standard area form. However, any pair of transverse foliations (and any area form) will do, so that even in this simple case the Künneth structure is not usually a global product.

In dimension 2 every almost Künneth structure is integrable, since every two-form is closed and every smooth line field is integrable. Every two-dimensional Künneth structure gives rise to a Lorentz metric. Conversely, every Lorentz metric on a surface determines a Künneth structure.

Here is a straightforward way of generating new examples from old ones.

Lemma 5.13 *Suppose $(\omega_1, \mathcal{F}_1, \mathcal{G}_1)$ and $(\omega_2, \mathcal{F}_2, \mathcal{G}_2,)$ are Künneth structures on M_1 and M_2 respectively. Then $(\omega_1 \pm \omega_2, \mathcal{F}_1 \times \mathcal{F}_2, \mathcal{G}_1 \times \mathcal{G}_2)$ are Künneth structures on $M_1 \times M_2$.*

Of course we can also consider $\mathcal{F}_1 \times \mathcal{G}_2$ and $\mathcal{F}_2 \times \mathcal{G}_1$ in place of $\mathcal{F}_1 \times \mathcal{F}_2$ and $\mathcal{G}_1 \times \mathcal{G}_2$.

Applying this lemma using for at least one of the factors a two-torus with a Künneth structure with complicated foliations, we see that in all even dimensions tori have non-standard, i. e. non-product, Künneth structures as well.

A natural and very interesting class of examples of Künneth structures arises on the cotangent bundles of affinely flat manifolds. Of course, this yields only non-compact examples.

Example 5.14 The cotangent bundle T^*M of any smooth manifold M carries the tautological exact symplectic form ω_{can}, and a Lagrangian foliation $\mathcal{F}$ whose leaves are the fibres of the projection $\pi \colon T^*M \longrightarrow M$. If we assume that M is affinely flat, meaning that TM carries a torsion-free flat connection

∇, then the proof of Theorem 4.13 shows that the horizontal distribution for ∇ is tangent to a Lagrangian foliation $\mathcal{G}$. Since this is complementary to $\mathcal{F}$, T^*M carries a Künneth structure whenever M is affinely flat.

In the next few subsections we discuss constructions that, unlike the previous example, allow us to also build Künneth structures on some compact manifolds that are not tori and not even finitely covered by tori.

5.3.1 Anosov Symplectomorphisms

Here we give the details for an important class of examples already mentioned in Subsection 1.1.3 of the introductory chapter.

Throughout, M will denote a closed smooth manifold. Here is the basic definition.

Definition 5.15 A diffeomorphism $\varphi\colon M \to M$ is called *Anosov* if there is a continuous splitting of the tangent bundle into invariant subbundles of positive rank $TM = E^s \oplus E^u$ such that for all $k > 0$

$$\|D\varphi^k(v)\| \le a \cdot e^{-bk}\|v\| \quad \forall v \in E^s\,,$$

$$\|D\varphi^k(v)\| \ge a \cdot e^{bk}\|v\| \quad \forall v \in E^u\,,$$

for some positive constants a and b.

Here the norms are taken with respect to some arbitrary Riemannian metric. There is the related concept of an expanding diffeomorphism, for which the second inequality holds for all tangent vectors. This is excluded here by the assumption that both E^s and E^u have positive rank.

Remark 5.16 While the precise values of the constants a and b depend on the choice of metric, the property of being Anosov does not. If the defining inequalities hold for some metric, then they hold for every metric (with different constants). One could also write the definitions using different constants in the two inequalities, but this seemingly more general definition is actually equivalent to the one above.

Remark 5.17 If φ is Anosov, so is φ^{-1}, with the rôles of E^s and E^u interchanged. (This explains how to rewrite the defining inequalities for negative k.)

The defining property of an Anosov diffeomorphism is sometimes referred to as the existence of an Anosov splitting $TM = E^s \oplus E^u$ into stable (or contracting) and unstable (or dilating) subbundles E^s and E^u respectively. This means that φ is hyperbolic everywhere. It is easy to see that when an Anosov splitting exists, it is uniquely determined by φ, as the contracting and dilating subspaces have to be maximal with these properties.

It is known that the subbundles E^s and E^u are actually tangent to foliations $\mathcal{F}$ and $\mathcal{G}$ with smooth leaves, although the subbundles are only assumed continuous. These foliations are called the stable and unstable foliations of φ. In general they are not smooth transversely to the leaves, but they are Hölder continuous. There are several interesting 'rigidity' results that show that Anosov diffeomorphisms with smooth Anosov splitting are quite special and can be described fairly explicitly.

Here is a result that was discussed informally in the introduction.

Proposition 5.18 *Let (M, ω) be a closed symplectic manifold admitting an Anosov symplectomorphism $\varphi \colon M \longrightarrow M$. Then E^s and E^u are Lagrangian with respect to ω. Therefore the stable and unstable foliations of φ together with the symplectic form ω define a Künneth structure.*

Proof Suppose $v, w \in E^s$. Then

$$\omega(v, w) = (\varphi^* \omega)(v, w) = \omega(D\varphi(v), D\varphi(w)) = \ldots = \omega(D\varphi^k(v), D\varphi^k(w)) .$$

Using the auxiliary metric g, we find that there is a constant c such that

$$|\omega(v, w)| \leq c \cdot \|\omega\| \cdot \|D\varphi^k(v)\| \cdot \|D\varphi^k(w)\| \leq c \cdot \|\omega\| \cdot a^2 \cdot e^{-2bk} \cdot \|v\| \cdot \|w\| .$$

Letting k go to infinity, the right-hand side becomes arbitrarily small. Therefore $\omega(v, w) = 0$ and E^s is ω-isotropic. By the same argument with φ^{-1} replacing φ we conclude that E^u is also ω-isotropic. As the two subbundles are complementary, they must be equidimensional and Lagrangian. □

In the situation of Proposition 5.18, the symplectomorphism φ is tautologically an automorphism of the Künneth structure it defines. The automorphism φ and its powers show that the automorphism group must be non-compact.

Corollary 5.19 *Every Anosov symplectomorphism of a closed symplectic manifold defines a Künneth structure with non-compact automorphism group.*

Here is the simplest example of an Anosov symplectomorphism.

Example 5.20 Consider $\varphi = \begin{pmatrix} 2 & 1 \\ 1 & 1 \end{pmatrix}$ acting as a linear diffeomorphism on $\mathbb{R}^2$. Since it preserves the integer lattice, it descends to a diffeomorphism of T^2. This diffeomorphism is area preserving because the determinant of the matrix is $= 1$. There are two real eigenvalues, one > 1 and one < 1, and their eigenspaces define the unstable and the stable subbundles, respectively, of the tangent bundle to the torus.

More generally, we can replace the given matrix by any hyperbolic element of $SL_2(\mathbb{Z})$. Similar examples exist on high-dimensional tori, and on other nil- and infra-nilmanifolds; see Section 9.6 of Chapter 9.

5.3.2 Suspensions

Before defining suspensions of arbitrary representations into Künneth auto- morphism groups, we consider the most useful special case. This is interesting in its own right and has generalisations other than suspensions; see Proposi- tion 5.27 below.

Whenever we have an automorphism $\varphi \in \mathrm{Aut}(\omega, \mathcal{F}, \mathcal{G})$ of a Künneth struc- ture, we can form a new Künneth structure as follows.

Proposition 5.21 *Let $\varphi \colon M \longrightarrow M$ be an automorphism of a Künneth struc- ture $(\omega, \mathcal{F}, \mathcal{G})$. Then the product $N = M(\varphi) \times S^1$, where $M(\varphi)$ is the mapping torus of φ, carries an induced Künneth structure.*

We will say that this Künneth structure is obtained by suspending φ.

Proof Recall that $M(\varphi)$ is the quotient of $M \times \mathbb{R}$ by the diagonal action of $\mathbb{Z} = \pi_1(S^1)$ given by (powers of) φ on M and by integer translations on $\mathbb{R}$. Therefore $M(\varphi)$ fibres over the circle, $\pi \colon M(\varphi) \longrightarrow S^1$. We denote by $\alpha = \pi^* d\theta$ the closed one-form on $M(\varphi)$ obtained as the pullback of a volume form on the circle. A volume form on the second factor in the product $N = M(\varphi) \times S^1$ will be denoted by dt. In general, forms on the two factors will be considered as forms on the product N, dropping the pullback under the projections from the notation.

Now because φ preserves ω, we can think of ω as a closed form of maximal rank on the mapping torus, which restricts to the original symplectic form on every fibre of π. The form $\Omega = \omega + \alpha \wedge dt$ is a symplectic form on N because it is clearly closed, and it is non-degenerate because $\omega^n \wedge \alpha$ is a volume form on $M(\varphi)$ if $\dim(M) = 2n$.

Since the automorphism φ preserves $\mathcal{F}$, the foliation $\mathcal{F} \times \mathbb{R}$ on $M \times \mathbb{R}$ de- scends to $M(\varphi)$ as a foliation $\widetilde{\mathcal{F}}$ of dimension $n + 1$. The automorphism φ also preserves $\mathcal{G}$ and therefore $\mathcal{G}$ defines an n-dimensional foliation on $M(\varphi)$. On the product $N = M(\varphi) \times S^1$ the foliation $\widetilde{\mathcal{G}} = \mathcal{G} \times S^1$ is complementary to $\widetilde{\mathcal{F}}$. Now we claim that these two foliations are Lagrangian, so that $(\Omega, \widetilde{\mathcal{F}}, \widetilde{\mathcal{G}})$ is indeed a Künneth structure.

Since the leaves of the foliation $\widetilde{\mathcal{F}}$ are contained in $M(\varphi)$, the form dt re- stricts trivially to them. Thus Ω and ω have the same restriction to $\widetilde{\mathcal{F}}$, and this vanishes since $\mathcal{F}$ is Lagrangian for ω. Similarly, the form α vanishes on the leaves of $\widetilde{G}$, and so Ω and ω have the same restriction to $\widetilde{G}$. Again this vanishes since G is Lagrangian for ω. We have verified that $\widetilde{\mathcal{F}}$ and $\widetilde{G}$ are Ω-isotropic. Since they are complementary, they are both Lagrangian. $\square$

If φ is isotopic to the identity of M, then the product $N = M(\varphi) \times S^1$, which is sometimes called a symplectic mapping torus, is diffeomorphic to $M \times T^2$. This of course has Künneth structures obtained by applying Lemma 5.13, so that in this case the more complicated suspension is not required. However, whenever φ is not isotopic to the identity, we do obtain new examples. For example, we can apply Proposition 5.21 to all Anosov symplectomorphisms, considered as automorphisms of their own Künneth structures via Proposition 5.18.

Example 5.22 Consider $M = T^2$ and an Anosov automorphism φ as in Example 5.20. Its suspension is a Künneth structure on a non-trivial T^2-bundle N over T^2. This N is a solvmanifold with first Betti number $b_1(N) = 2$.

This is the first example of a Künneth structure on a closed manifold that is not an infra-nilmanifold.

We now come to the general suspension construction.

Theorem 5.23 *Let B and F be smooth manifolds endowed with Künneth structures $(\omega_B, \mathcal{F}_B, G_B)$ and $(\omega_F, \mathcal{F}_F, G_F)$. For every representation*

$$\rho \colon \pi_1(B) \longrightarrow \mathrm{Aut}(\omega_F, \mathcal{F}_F, G_F) \subset \mathrm{Diff}_+(F)$$

the total space of the corresponding flat F-bundle $N_\rho \longrightarrow B$ carries an induced Künneth structure that is locally the product of $(\omega_B, \mathcal{F}_B, G_B)$ and $(\omega_F, \mathcal{F}_F, G_F)$.

Proof The Künneth structure $(\omega_B, \mathcal{F}_B, G_B)$ lifts to a Künneth structure on the universal covering $\widetilde{B}$, which is invariant under deck transformations. By Lemma 5.13 this lifted structure together with the Künneth structure on F gives a product Künneth structure on $\widetilde{B} \times F$. We consider the diagonal action of the fundamental group $\pi_1(B)$ on $\widetilde{B} \times F$, where the action on the first factor is by deck transformations and the action on the second factor is via ρ. This group action is by automorphisms of the product Künneth structure, and so this structure descends to one on the flat F-bundle

$$(\widetilde{B} \times F)/\pi_1(B) = N_\rho \longrightarrow B \,.$$

This completes the proof. $\square$

Clearly the construction of Proposition 5.21 corresponds to the special case of this theorem with $B = T^2$, $F = M$, and the representation

$$\rho \colon \pi_1(T^2) = \mathbb{Z}^2 \longrightarrow \mathrm{Aut}(\omega, \mathcal{F}, G)$$

sending one standard generator of $\mathbb{Z}^2$ to φ and the other to Id_M. In the proof of Proposition 5.21 we implicitly used a trivial Künneth structure on the base T^2. The proof of Theorem 5.23 shows that we could instead use any Künneth structure on T^2.

Here is a consequence of Theorem 5.23 that uses pairs of automorphisms of different Künneth structures to generate new examples on products of mapping tori.

Corollary 5.24 *Let φ_i, for $i = 1, 2$, be automorphisms of Künneth structures $(\mathcal{F}_i, \mathcal{G}_i, \omega_i)$ defined on M_i. Then the product $M(\varphi_1) \times M(\varphi_2)$ of their mapping tori carries an induced Künneth structure.*

Proof We may think of $\varphi_1 \times \mathrm{Id}_{M_2}$ and $\mathrm{Id}_{M_1} \times \varphi_2$ as automorphisms of the product Künneth structure on $M_1 \times M_2$. As the two automorphisms commute with each other, we can define a representation ρ of $\mathbb{Z}^2$ into the automorphisms of the product Künneth structure by sending the first generator to $\varphi_1 \times \mathrm{Id}_{M_2}$ and the second generator to $\mathrm{Id}_{M_1} \times \varphi_2$. Suspending this ρ we obtain a Künneth structure on the total space of the corresponding $(M_1 \times M_2)$-bundle over T^2. This total space is diffeomorphic, and isomorphic as a flat bundle, to the product $M(\varphi_1) \times M(\varphi_2)$ of mapping tori. $\square$

Remark 5.25 One may think of Proposition 5.21 as the special case of Corollary 5.24 in which M_2 is a point.

Remark 5.26 The same proof works for an arbitrarily large number $\varphi_1, \ldots, \varphi_k$ of automorphisms of (potentially) different Künneth structures, showing that if k is even, then the product $M(\varphi_1) \times \ldots \times M(\varphi_k)$ of mapping tori carries a Künneth structure. If k is odd, one just adds a copy of a point and considers $M(\varphi_1) \times \ldots \times M(\varphi_k) \times S^1$ instead.

5.3.3 A Variation on the Theme of Suspensions

A variation on the suspension construction is obtained starting from Proposition 5.21 and relaxing the condition that φ preserve the full Künneth structure. We will consider area-preserving diffeomorphisms of the two-torus, not necessarily hyperbolic, but with the property that they preserve a one-dimensional foliation.

Proposition 5.27 *Assume that the action of $\varphi \in SL_2(\mathbb{Z})$ as a linear diffeomorphism of T^2 preserves a one-dimensional foliation $\mathcal{F}$. Then the four-manifold $N = M(\varphi) \times S^1$ carries an induced Künneth structure.*

Proof We adapt the proof of Proposition 5.21. The diffeomorphism $\varphi \colon T^2 \longrightarrow T^2$ preserves the standard area form $dx \wedge dy$, so this descends to a closed two-form ω of maximal rank on the mapping torus. As before, we denote by α the closed one-form on $M(\varphi)$ obtained as the pullback of a volume form on the circle, so that $\omega \wedge \alpha$ is a volume form on $M(\varphi)$.

Since φ preserves $\mathcal{F}$, this can be suspended to yield a two-dimensional foliation $\widetilde{\mathcal{F}}$ on the mapping torus. Although there is no natural complement preserved by φ, we can choose a one-dimensional foliation $\mathcal{G}$ on $M(\varphi)$ that is complementary to $\widetilde{\mathcal{F}}$ and everywhere tangent to the fibres of the projection $\pi \colon M(\varphi) \longrightarrow S^1$. This can be done for example by choosing an arbitrary Riemannian metric on $M(\varphi)$ and considering the orthogonal complement of $\mathcal{F}$ in the restrictions of the metric to the fibres of the mapping torus. Integrability is automatic because the subbundle in question is of rank 1.

In this situation $\widetilde{\mathcal{G}} = \mathcal{G} \times S^1$ is complementary to $\widetilde{\mathcal{F}}$ on N. As in the proof of Proposition 5.21, the foliations $\widetilde{\mathcal{F}}$ and $\widetilde{\mathcal{G}}$ are Lagrangian for $\Omega = \omega + \alpha \wedge dt$. The arbitrariness in the choice of $\mathcal{G}$ does not spoil the argument because every one-dimensional foliation on T^2 is Lagrangian. $\qquad\square$

Here is an example of a non-hyperbolic φ to which this construction applies.

Example 5.28 We may take $\varphi = \begin{pmatrix} 1 & 1 \\ 0 & 1 \end{pmatrix}$. This preserves one of the two foliations of T^2 by its factors, giving rise to an invariant foliation $\mathcal{F}$. The other foliation by factors is not preserved, but we can still find a $\mathcal{G}$ as above, tangent to the fibres of the mapping torus and complementary to $\mathcal{F}$ on every fibre.

Here φ is a Dehn twist along one of the circle factors of T^2, and its mapping torus $M(\varphi)$ is a nilmanifold. The product $N = M(\varphi) \times S^1$ is the Kodaira–Thurston manifold. It has first Betti number 3.

It should be clear from the examples we have given so far that constructing Künneth structures is quite a difficult and delicate matter. While symplectic structures have no local invariants and are therefore quite flexible and can be modified by certain cut-and-paste operations, Künneth structures do have local geometry and are hard to manipulate because, while modifying the symplectic form, one also has to keep control of the Lagrangian subbundles and their integrability. If we give up integrability, we are left with only almost Künneth structures, for which existence can be discussed purely in terms of algebraic topology.

5.4 Remarks on Terminology

We have defined a Künneth structure as a bifoliation equipped with a symplectic structure of type $(1, 1)$. Equivalent definitions have appeared many times in the literature, under many different names.

As far as we know, the first to consider these structures was Raševskiĭ in 1948 (see [Ras-48]), and several authors from the Soviet Union immediately followed up on his work. In the Western literature, Libermann [Lib-52, Lib-54] was the first to consider bi-Lagrangian structures, under the name of **para-Kähler structures**. Her work was in the context of para-complex geometry, in which one does geometry over the para-complex numbers instead of the complex numbers. The para-complex numbers are the $\mathbb{R}$-algebra $\mathbb{D}$ generated by 1 and I, where $I^2 = 1$. There is an enormous amount of literature on para-complex geometry, and we refer to the surveys [CFG-96, AMT-09] for more references. We give a precise definition of para-Kähler structures in Section 6.3 of Chapter 6, where we prove the equivalence between para-Kähler and bi-Lagrangian structures.

The algebra $\mathbb{D}$ appears in the literature under many alternative names, and each of these names is then propagated in the names of geometric structures based on $\mathbb{D}$.

One of these alternative names for $\mathbb{D}$ is the hyperbolic numbers. Based on this there is a certain amount of literature discussing hyperbolic Hermitian manifolds. In this case there is a notion of **hyperbolic Kähler structures**, and these are trivially seen to be the same as a para-Kähler structures. The literature referred to in [CFG-96, AMT-09] switches easily between the hyperbolic Kähler and the para-Kähler terminologies.

The algebra $\mathbb{D}$ is sometimes called the double numbers, the term preferred in [HL-12]. This gives rise to double manifolds or $\mathbb{D}$-manifolds, which are manifolds equipped with an equidimensional bifoliation. Again there are notions of Hermitian and Kähler structures in this category, and they match the terminology of para-complex geometry. Therefore, **Kähler $\mathbb{D}$-structures** in the terminology of [HL-12] are nothing but para-Kähler structures, which in turn are the same as bi-Lagrangian or Künneth structures.

One could also call $\mathbb{D}$ the split complex numbers, or the split-form analogue of the complex numbers. Although this name is not very common, the corresponding name of split quaternions for the split form analogue of the algebra of quaternions is quite common; see for example [DJS-08]. In any case, although the name split complex numbers is not used much for $\mathbb{D}$, many authors do describe bi-Lagrangian geometry as a split-form analogue of Kähler geometry; see for example Bryant [Bry-01, Section 5.2].

In his work on geodesic flows of negatively curved manifolds, Kanai [Kan-88] encountered symplectic manifolds with Lagrangian bifoliations, and introduced the name **bipolarised symplectic manifolds** for them. This name is derived from the name polarisation, which is sometimes used for Lagrangian foliations, particularly in the context of geometric quantisation. The situation that Kanai studied arises from a continuous time analogue of the Anosov symplectomorphisms discussed in Subsection 5.3.1 above, because for manifolds of strictly negative sectional curvature, the geodesic flow on the unit tangent bundle is a contact Anosov flow.

Tabachnikov [Tab-93] used the name **Lagrangian two-webs** for bi-Lagrangian structures, making contact with classical web geometry in the style of Blaschke and Chern.

In December of the year 2000, the second author attended a seminar talk at Harvard with the title *Künneth geometry*, given by Victor Guillemin. In this talk [Gui-00], Guillemin first discussed a problem in Kähler geometry, and then went on to modify that problem, by replacing the complex structure of a Kähler manifold by a local product structure, or bifoliation. He noted that a local product structure gives rise to a bigrading on differential forms, and he defined Künneth structures the way we have defined them at the beginning of this chapter. So a 'Künneth structure' is just a new name for bi-Lagrangian, or bipolarized, or para-Kähler, or hyperbolic Kähler, or split Kähler structures.

There are several reasons why we have adopted Guillemin's terminology from [Gui-00], and are calling these structures Künneth structures, rather than using one of the older names. One is that this was the name under which we first learned about these structures, and after a while we became attached to the name and the ensuing terminology. Another is that we did not want to choose sides and join one of the schools using one of the older names to the exclusion of the others. Finally, we know from our own experience that specialised terminologies like 'para-...' or 'split' often signal niche topics that are somewhat outside the mainstream and come with their own jargon, which may deter the uninitiated. We have tried to keep our treatment free of such jargon, and we want to make the case for Künneth geometry as a mainstream topic of differential topology unencumbered by specialised notions that are analogues of familiar structures in an unfamiliar setting. So, rather than treating Künneth geometry as the lesser known sibling of some well-known geometry, we study it in its own right.

Notes for Chapter 5

1. We learned Proposition 5.3 from Guillemin [Gui-00]. It also appears in papers of Tabachnikov [Tab-93], of Cortés, Mayer, Mohaupt and Saueressig [CMMS-04, Theorem 2], and perhaps of others.

2. Theorem 5.12 is at least implicit in Gromov's theory [Gro-88] of rigid geometric structures, and is made explicit in [DG-91, Section 2.6]. Its local version, pertaining to the pseudogroup of local diffeomorphisms preserving the structure, is what allows one to apply Gromov's theory to Künneth structures, as is done in [BFL-92, BL-93].

3. Example 5.14 is essentially due to Weinstein [Wei-71]. It also appears in Bejan [Bej-94].

4. Proposition 5.18 is well known. The earliest reference we have been able to locate is [DG-91, Section 2.6]. The analogous statement for contact Anosov flows appears in Kanai [Kan-88, Lemma (2.2)].

6

The Künneth Connection

In this chapter we discuss the unique torsion-free affine connection defined by a Künneth structure. This connection, which we call the Künneth connection, preserves the two foliations and is compatible with the symplectic structure. We will actually start with a more general setup, proving that for every almost Künneth structure there is a distinguished connection for which the whole structure is parallel. It then turns out that this connection is torsion free if and only if the almost Künneth structure is integrable, i.e. it arises tautologically from a Künneth structure. In this case the Künneth connection is just the Levi–Civita connection of the associated pseudo-Riemannian metric.

In the final section of this chapter we use connections to prove that Künneth or bi-Lagrangian structures are in fact the same as para-Kähler structures.

6.1 The Canonical Connection of an Almost Künneth Structure

Let (ω, F, G) be an almost Künneth structure on a smooth manifold M, so that ω is a non-degenerate two-form and F and G are complementary Lagrangian subbundles of (TM, ω). We are interested in connections ∇ on the tangent bundle $TM \longrightarrow M$ with the following properties:

(i) ∇ preserves both F and G, i.e. parallel transport with respect to ∇ preserves the splitting $TM = F \oplus G$,

(ii) ∇ is compatible with ω, meaning that

$$L_X(\omega(Y, Z)) = \omega(\nabla_X Y, Z) + \omega(Y, \nabla_X Z)$$

for all vector fields X, Y and Z.

We discussed this second property only for symplectic forms ω in Chapter 4 (see Definition 4.18), but of course it makes sense in the almost symplectic setting.

The requirements above leave the torsion T^∇ of the connection ∇ unspecified. Note that according to Proposition 4.17 we should not expect that the torsion of ∇ vanishes, because this would imply that F and G are integrable, which is something we do not assume – and which is often false! – for an almost Künneth structure. We will require only the vanishing of certain special components of the torsion, defined as follows.

Definition 6.1   The *mixed torsion* of ∇ consist of $T^\nabla(X, Y)$ where $X \in F$ and $Y \in G$ or vice versa, and T^∇ is the usual torsion tensor of ∇ defined by

$$T^\nabla(X, Y) = \nabla_X Y - \nabla_Y X - [X, Y] .$$

We will prove that the almost Künneth structure admits a unique affine connection satisfying (i) and (ii) above, and so that the mixed torsion vanishes. For the existence we give a direct construction, on which we now embark. The following definition is motivated by the formula (4.1) for the Bott connection of a Lagrangian foliation.

Definition 6.2   Let X, Y be arbitrary vector fields on M. Since ω is non-degenerate, there is a unique vector field $D(X, Y)$ satisfying

$$i_{D(X,Y)}\omega = L_X i_Y \omega .$$

It is clear that the map

$$D: \mathfrak{X}(M) \times \mathfrak{X}(M) \longrightarrow \mathfrak{X}(M)$$

is $\mathbb{R}$-linear in both entries. In order to check its suitability as a (partial) connection, we investigate how it behaves when we multiply the vector fields by functions $f \in C^\infty(M)$.

Lemma 6.3   *The operator D satisfies the Leibniz rule*

$$D(X, fY) = fD(X, Y) + (L_X f)Y .$$

Proof   Using just the definition of D and the product rule for the Lie derivative L_X we find

$$\begin{aligned}
i_{D(X,fY)}\omega &= L_X i_{fY}\omega \\
&= L_X(f i_Y \omega) \\
&= (L_X f)i_Y \omega + f L_X i_Y \omega \\
&= i_{(L_X f)Y}\omega + i_{fD(X,Y)}\omega \\
&= i_{(L_X f)Y + fD(X,Y)}\omega .
\end{aligned}$$

The claim now follows from the non-degeneracy of ω. $\qquad\square$

Lemma 6.4 *If the vector fields X and Y satisfy $\omega(X,Y) = 0$, then we have $D(fX,Y) = fD(X,Y)$.*

Proof On the one hand, we have

$$i_{D(fX,Y)}\omega = L_{fX}i_Y\omega = di_{fX}i_Y\omega + i_{fX}di_Y\omega = fi_Xdi_Y\omega,$$

where the first equality is the definition of D, the second is the Cartan formula and the third is true because $\omega(Y,fX) = 0$.

On the other hand, we have

$$i_{fD(X,Y)}\omega = fi_{D(X,Y)}\omega = fL_Xi_Y\omega = fdi_Xi_Y\omega + fi_Xdi_Y\omega = fi_Xdi_Y\omega,$$

where the first equality is the $C^\infty(M)$-linearity of the contraction, the second is the definition of D, the third is the Cartan formula, and the last one is true because $\omega(Y,X) = 0$. This proves the claim. $\qquad\square$

The Whitney sum decomposition $TM = F \oplus G$ allows us to uniquely decompose any vector field X on M into its components X_F and X_G contained in F and in G, respectively.

Proposition 6.5 *Let X be any vector field on M, and Y a section of F. Then*

$$\nabla_X Y = D(X_F, Y)_F + [X_G, Y]_F \tag{6.1}$$

defines a connection on the vector bundle $F \longrightarrow M$.

Proof By definition, ∇ is $\mathbb{R}$-linear. We need to check that it is $C^\infty(M)$-linear in X and satisfies the Leibniz rule. For the first point we compute

$$
\begin{aligned}
\nabla_{fX} Y &= D(fX_F, Y)_F + [fX_G, Y]_F \\
&= fD(X_F, Y)_F + (f[X_G, Y] - (L_Y f)X_G)_F \\
&= f(D(X_F, Y) + [X_G, Y])_F \\
&= f\nabla_X Y,
\end{aligned}
$$

where the first equality is the definition of ∇, the second is true by Lemma 6.4, the third is true because $(X_G)_F = 0$ for all X, and the last one is again the definition of ∇.

For the Leibniz rule we need only the definition of ∇ and Lemma 6.3:

$$
\begin{aligned}
\nabla_X(fY) &= D(X_F, fY)_F + [X_G, fY]_F \\
&= fD(X_F, Y)_F + (L_{X_F}f)Y + f[X_G, Y]_F + (L_{X_G}f)Y \\
&= f\nabla_X Y + (L_X f)Y.
\end{aligned}
$$

This completes the proof that ∇ is a connection on the vector bundle F. $\qquad\square$

In the same way, we can define a connection on the vector bundle G over M by the formula

$$\nabla_X Y = D(X_G, Y)_G + [X_F, Y]_G, \tag{6.2}$$

where X is any vector field on M and Y is a section of G.

Forming the direct sum connection of the two connections we have defined on F and G will prove the existence part of the following result.

Theorem 6.6 *Let (ω, F, G) be an almost Künneth structure on a smooth manifold M. There exists a unique affine connection on M that preserves both F and G, that is compatible with ω and whose mixed torsion vanishes identically.*

Proof We consider the direct sum connection on $TM \longrightarrow M$, given by

$$\nabla_X Y = \nabla_X Y_F + \nabla_X Y_G,$$

where the terms on the right-hand side are defined using (6.1) for F and by (6.2) for G. This direct sum connection will turn out to be the unique connection characterised by the theorem, and so we will call it the *Künneth connection* of the structure (ω, F, G).

By construction, the Künneth connection preserves both subbundles F and G. We now prove that it is also compatible with ω. For this we have to show that

$$L_X(\omega(Y,Z)) = \omega(\nabla_X Y, Z) + \omega(Y, \nabla_X Z) \tag{6.3}$$

for all vector fields X, Y and Z on M. We show that the two sides of the equation are equal for

$$Y, Z \in \Gamma(F), \quad X \in \mathfrak{X}(M)$$

and for

$$X, Y \in \Gamma(F), \quad Z \in \Gamma(G).$$

By the same argument this also follows if we interchange the rôles of F and G. Together with $\mathbb{R}$-linearity, this shows that formula (6.3) holds for all $X, Y, Z \in \mathfrak{X}(M)$.

As the subbundle F is isotropic with respect to ω and is preserved by the Künneth connection, both sides of equation (6.3) vanish if Y and Z are sections of F.

Suppose X and Y are sections of F, and Z is a section of G. Then

$$\omega(\nabla_X Y, Z) + \omega(Y, \nabla_X Z) = \omega(D(X, Y)_F, Z) + \omega(Y, [X, Z]_G)$$
$$= \omega(D(X, Y), Z) + \omega(Y, [X, Z])$$
$$= (L_X i_Y \omega)(Z) + \omega(Y, [X, Z])$$
$$= L_X(i_Y \omega(Z)) - (i_Y \omega)(L_X Z) + \omega(Y, [X, Z])$$
$$= L_X(i_Y \omega(Z)) - \omega(Y, [X, Z]) + \omega(Y, [X, Z])$$
$$= L_X(i_Y \omega(Z))$$
$$= L_X(\omega(Y, Z)).$$

In the second line of the sequence of equations, we used that F and G are isotropic with respect to ω.

We have now completed the verification that our direct sum connection is compatible with ω. Finally we calculate its mixed torsion, and this will finish the existence part of the proof of Theorem 6.6. The torsion tensor

$$T^\nabla(X, Y) = \nabla_X Y - \nabla_Y X - [X, Y]$$

is skew-symmetric, and so it suffices to consider the case that X is a section of F and Y is a section of G. Then $X_G = 0 = Y_F$, and so

$$T^\nabla(X, Y) = [X, Y]_G - [Y, X]_F - [X, Y] = 0,$$

showing that the mixed torsion vanishes.

It remains to prove the uniqueness part of Theorem 6.6. For this let ∇' be any connection that preserves the subbundles F and G, that is compatible with ω and such that its mixed torsion vanishes. We need to show that ∇' agrees with the Künneth connection ∇ discussed above.

Let X be a section of F and Y a section of G. Then $T^{\nabla'}(X, Y) = 0$ and therefore

$$\nabla'_X Y = (\nabla'_X Y)_G = (\nabla'_Y X + [X, Y])_G = [X, Y]_G = \nabla_X Y,$$

because $\nabla'_Y X$ is a section of F.

Now assume that both X and Y are sections of F. By definition of the map D we have for an arbitrary vector field Z

$$\omega(D(X, Y), Z) = (L_X i_Y \omega)(Z) = L_X(\omega(Y, Z)) - \omega(Y, [X, Z]). \qquad (6.4)$$

If Z is a section of G, then, because both subbundles are isotropic with respect to ω, we find

$$
\begin{aligned}
\omega(D(X,Y)_F, Z) &= \omega(D(X,Y), Z) \\
&= L_X(\omega(Y,Z)) - \omega(Y, [X,Z]) \\
&= L_X(\omega(Y,Z)) - \omega(Y, [X,Z]_G) \\
&= L_X(\omega(Y,Z)) - \omega(Y, \nabla'_X Z) \\
&= \omega(\nabla'_X Y, Z) \,,
\end{aligned}
$$

where we have used (6.4) and the assumption that ∇' is compatible with ω. Since $\nabla'_X Y$ is a section of F, the last equation actually also holds for $Z \in F$ and thus for all $Z \in TM$, proving that

$$
\nabla'_X Y = D(X,Y)_F = \nabla_X Y \,.
$$

Using the same argument with the rôles of F and G reversed proves that ∇' is equal to the Künneth connection ∇. This completes the proof of Theorem 6.6.

$\square$

The following example might look rather trivial at first sight, but it has interesting consequences; see Subsection 7.2.1 of Chapter 7.

Example 6.7   Suppose (ω_1, F_1, G_1) and (ω_2, F_2, G_2) are almost Künneth structures on M_1 and M_2 respectively. Then $(\omega_1 \pm \omega_2, F_1 \oplus F_2, G_1 \oplus G_2)$ are almost Künneth structures on $M_1 \times M_2$, whose Künneth connections are the direct sum connections of the Künneth connections on the two factors.

6.2 The Torsion of the Künneth Connection

By definition, the mixed torsion of the Künneth connection vanishes for any almost Künneth structure. The remaining torsion components precisely measure the non-integrability of the structure. This is of course not surprising in light of the discussion in Section 4.3 of Chapter 4.

Theorem 6.8   *An almost Künneth structure is integrable, i.e. arises tautologically from a Künneth structure, if and only if its Künneth connection is torsion free.*

Proof   In one direction, if the Künneth connection ∇ of (ω, F, G) is torsion free, then since it preserves F and G, these subbundles are integrable by Proposition 4.17. They are therefore tangent to foliations, which are of course Lagrangian for ω, because F and G are. Moreover, since ∇ is compatible with ω, one easily computes

$$d\omega(X, Y, Z) = \omega(T^{\nabla}(X, Y), Z) - \omega(T^{\nabla}(X, Z), Y) + \omega(T^{\nabla}(Y, Z), X) \, . \quad (6.5)$$

This shows that if ∇ is torsion free, then ω is closed. This is a particular case of the general principle that parallel forms for torsion-free connections are closed; see the proof of Theorem 4.13.

Conversely, let us assume that the almost Künneth structure is integrable, so that F and G are tangent to Lagrangian foliations $\mathcal{F}$ and $\mathcal{G}$, and $d\omega \equiv 0$. Take sections X and Y of $F = T\mathcal{F}$. Since this subbundle is integrable and preserved by ∇, we conclude that $T^{\nabla}(X, Y)$ takes values in F. As F is Lagrangian, $\omega(T^{\nabla}(X, Y), Z) = 0$ whenever Z is in F. Now if we take Z in G and look at (6.5), we see that the second and third summands on the right-hand side contain mixed torsion terms and therefore vanish. As the left-hand side also vanishes, we conclude $\omega(T^{\nabla}(X, Y), Z) = 0$ in this case as well. By $\mathbb{R}$-linearity we obtain the vanishing of this term for arbitrary Z, and so by the non-degeneracy of ω we conclude $T^{\nabla}(X, Y) = 0$ for all X and Y in F.

Reversing the rôles of F and G, this argument tells us that $T^{\nabla}(X, Y) = 0$ also when both X and Y are in G. Since the mixed torsion vanishes in any case, this shows that ∇ is torsion free. □

Corollary 6.9 *Let $(\omega, \mathcal{F}, \mathcal{G})$ be a Künneth structure on a smooth manifold M. Then its Künneth connection is the unique torsion-free affine connection on M that preserves the subbundles $T\mathcal{F}$ and $T\mathcal{G}$ and is compatible with the symplectic form ω.*

Proof According to Theorem 6.6 and Theorem 6.8 the Künneth connection has all the claimed properties. Conversely, if an affine connection has these properties, then it must be the Künneth connection according to the uniqueness part of Theorem 6.6 (which holds under the weaker assumption that the mixed torsion components vanish). □

This corollary bears some similarity to the characterisation of the Levi–Civita connection of a pseudo-Riemannian metric, and it suggests that for Künneth (rather than almost Künneth) structures the Künneth connection might agree with the Levi–Civita connection of the associated neutral metric g introduced in Proposition 5.9. This is indeed the case.

Theorem 6.10 *If $(\omega, \mathcal{F}, \mathcal{G})$ is a Künneth structure, then the Levi–Civita connection of the associated neutral metric g is the Künneth connection.*

Proof For the purposes of this proof we denote the Levi–Civita connection by ∇, although elsewhere this notation is often reserved for the Künneth connection. Note that although the metric g is well defined only up to an overall choice of sign, ∇ is well defined without any ambiguity, since both choices of sign for g give the same Levi–Civita connection.

By definition the Levi–Civita connection is torsion free. Therefore, we need to prove only that it is compatible with ω and preserves the subbundles $T\mathcal{F}$ and $T\mathcal{G}$, for the conclusion then follows from the uniqueness statement in Corollary 6.9.

We first check that ∇ preserves the subbundle $T\mathcal{F} \subset TM$. Since $T\mathcal{F}$ is maximally isotropic for ω, we have only to prove

$$\omega(Z, \nabla_Y X) = 0$$

for all $X, Z \in T\mathcal{F}$ and arbitrary $Y \in TM$. The definition of g in Section 5.2 of Chapter 5 together with the assumption that Z is in the $+1$-eigenspace of the almost product structure I gives

$$\omega(Z, \nabla_Y X) = \omega(IZ, \nabla_Y X) = g(Z, \nabla_Y X) \,,$$

and twice the right-hand side can now be written using the Koszul formula for ∇ to obtain

$$2g(Z, \nabla_Y X) = L_Y g(X, Z) + L_X g(Y, Z) - L_Z g(Y, X)$$
$$+ g([Y, X], Z) + g([Z, Y], X) + g(Y, [Z, X]) \,.$$

In this sum each scalar product has an argument that is X, Z, or $[Z, X]$, all of which are in the $+1$-eigenspace of I, by assumption for X and Z and because of the integrability of $T\mathcal{F}$ for the commutator. Therefore we can rewrite all these terms directly in terms of ω, and we can drop the first one because $T\mathcal{F}$ is ω-isotropic. This leads to

$$2\omega(Z, \nabla_Y X) = - L_X \omega(Y, Z) + L_Z \omega(Y, X)$$
$$- \omega([Y, X], Z) - \omega([Z, Y], X) - \omega(Y, [Z, X]) \,.$$

This sum vanishes, since, using $\omega(X, Z) = 0$ again, the sum equals $-d\omega(X, Y, Z)$ and ω is closed. This completes the proof that ∇ preserves $T\mathcal{F}$. The proof for $T\mathcal{G}$ is the same.

It remains to prove that ∇ is compatible with ω, in other words

$$L_X(\omega(Y, Z)) = \omega(\nabla_X Y, Z) + \omega(Y, \nabla_X Z)$$

for all tangent vectors X, Y and Z. If Y and Z are either both tangent to $\mathcal{F}$ or both tangent to $\mathcal{G}$, then both sides of the equation vanish because ∇ preserves $T\mathcal{F}$ and $T\mathcal{G}$. Therefore it is enough to check this equation for $Y \in T\mathcal{F}$ and

$Z \in T\mathcal{G}$. This means $IY = Y$ and $IZ = -Z$. We can now compute explicitly using the fact that the Levi–Civita connection ∇ is compatible with g:

$$
\begin{aligned}
L_X(\omega(Y, Z)) &= L_X \omega(IY, Z) = L_X g(Y, Z) \\
&= g(\nabla_X Y, Z) + g(Y, \nabla_X Z) \\
&= -g(\nabla_X Y, IZ) + g(IY, \nabla_X Z) \\
&= -\omega(Z, \nabla_X Y) + \omega(Y, \nabla_X Z) \\
&= \omega(\nabla_X Y, Z) + \omega(Y, \nabla_X Z) \,.
\end{aligned}
$$

This completes the proof of the theorem. □

Clearly the conclusion of Theorem 6.10 is false for non-integrable almost Künneth structures, since for them the Künneth connection must have torsion and so is not the Levi–Civita connection of any metric.

Remark 6.11 Note that according to Proposition 4.20 the Künneth connection of a Künneth (rather than an almost Künneth) structure restricts to the Bott connection along the leaves of $\mathcal{F}$ and $\mathcal{G}$. This can also be checked from the definition of the map D.)

Remark 6.12 Let (ω, F, G) be an almost Künneth structure, and suppose that F is integrable. Then the definition of the Künneth connection simplifies, because we have $D(X, Y) \in \Gamma(F)$ for sections $X, Y \in \Gamma(F)$, i.e. we do not need the projection onto the F-summand in the first term on the right-hand side of (6.1). To see this, note that F is maximally isotropic and, because of isotropy and integrability,

$$
\omega(D(X, Y), Z) = L_X(\omega(Y, Z)) - \omega(Y, [X, Z]) = 0 \,,
$$

for all sections Z of F.

6.3 Para-Kähler Structures

Para-complex geometry is an analogue of complex geometry, in which one replaces the complex numbers by the $\mathbb{R}$-algebra generated by an element I with $I^2 = 1$. The elements of this algebra are called para-complex or double numbers. We refer to [CFG-96] and the references therein for a survey of para-complex geometry.

An important topic in para-complex geometry is the so-called para-Kähler structures. We can now show that Künneth structures are in fact equivalent to para-Kähler structures. While we prefer the terminology of symplectic geometry and the geometry of foliations, this section is intended to show that all the contents of this book can be translated into the language and terminology of para-complex geometry. That language does not appear elsewhere in this book.

Here is the starting definition.

Definition 6.13 An *almost para-Hermitian structure* on a smooth manifold M consists of an almost product structure I and a pseudo-Riemannian metric g such that

$$g(IX, IY) = -g(X, Y) \quad \forall X, Y \in TM . \tag{6.6}$$

Since g is non-degenerate and symmetric, there must be vectors X with $g(X, X) \neq 0$. The equation (6.6) then shows that X and IX are g-orthogonal and span a two-dimensional subspace on which g has signature $(1, 1)$. Considering the orthogonal complement of this subspace and proceeding by induction, one concludes that $\dim(M) = 2n$ for some n and the signature of g is (n, n). Thus, although the signature of g was not explicitly mentioned in the definition, it has to be neutral.

For any almost para-Hermitian structure (I, g) we define ω by

$$\omega(X, Y) = g(IX, Y) .$$

This ω is skew-symmetric because $g(IX, IY) = -g(X, Y)$, and is non-degenerate because I is an isomorphism and g is non-degenerate. So ω is a non-degenerate two-form, called the fundamental two-form of the almost para-Hermitian structure.

We can now prove the following equivalence between almost para-Hermitian and almost Künneth structures.

Lemma 6.14 *Suppose that (I, g) is an almost para-Hermitian structure, and $F, G \subset TM$ are the eigenbundles of I. Then (ω, F, G) is an almost Künneth structure. Conversely, given an almost Künneth structure, its associated almost product structure I and neutral metric g form an almost para-Hermitian structure.*

Proof Suppose that (I, g) is an almost para-Hermitian structure. By (6.6), the eigenbundles F and G of I are null for g and therefore isotropic for ω. As they are complementary, it follows that (ω, F, G) is an almost Künneth structure.

Conversely, suppose we have some almost Künneth structure (ω, F, G) with associated almost product structure I and neutral metric g defined by $g(X, Y) =$

$\omega(IX, Y)$. Then

$$
\begin{aligned}
g(IX, IY) &= \omega(X, IY) \\
&= \omega(X_F + X_G, Y_F - Y_G) \\
&= -\omega(X_F, Y_G) + \omega(X_G, Y_F) \\
&= -\omega(X_F - X_G, Y_F + Y_G) \\
&= -\omega(IX, Y) \\
&= -g(X, Y) \,,
\end{aligned}
$$

and so (I, g) is an almost para-Hermitian structure.   □

The two procedures employed in the proof are clearly inverses of each other if one remembers that, in one direction, an almost Künneth structure determines I and g only up to an overall sign, and, in the other direction, one has to make a choice of which eigenbundle to call F and which to call G. We therefore obtain an equivalence between almost Künneth structures and almost para-Hermitian structures up to sign.

We can compare Definition 6.13 to the definition of an almost pseudo-Hermitian structure, in which one requires an almost complex structure J instead of the almost product structure I, together with a pseudo-Riemannian metric g satisfying

$$
g(JX, JY) = g(X, Y) \quad \forall X, Y \in TM \,.
$$

In this case, the definition does not determine the signature of g, and if one wants to recover the standard notion of an almost Hermitian structure, then one has to make positive-definiteness of the metric g a part of the definition.

For almost Hermitian structures there are two different integrability conditions one can impose. One is the integrability of J, giving rise to a complex structure with a Hermitian metric. The other, which is independent of the first, is the condition that the fundamental two-form of the almost Hermitian structure be closed, giving rise to an almost Kähler structure. If one adds to the definition of an almost Hermitian structure the requirement that J commute with the Levi–Civita connection of g, then both of these integrability conditions are satisfied, because an almost complex structure that commutes with a torsion-free connection must be integrable, and if J commutes with the Levi–Civita connection then the fundamental two-form is parallel and hence closed. In the case that J commutes with the Levi–Civita connection of g, the almost Hermitian structure becomes a Kähler structure.

This motivates the following definition.

Definition 6.15   An almost para-Hermitian structure (I, g) on a smooth manifold M is called *para-Kähler* if I commutes with the Levi–Civita connection of g.

We can now extend Lemma 6.14 to the integrable case.

Theorem 6.16   *An almost para-Hermitian structure is para-Kähler if and only if the corresponding almost Künneth structure is Künneth.*

Proof   In one direction, if I commutes with the Levi–Civita connection ∇, then by Proposition 4.25 its eigenbundles are integrable to a bifoliation. Moreover, since g is parallel for ∇, the definition of the fundamental two-form ω implies that it is parallel as well. In particular, it is closed since parallel forms for torsion-free connections are closed. Therefore ω together with the Lagrangian bifoliation obtained by integrating I is a Künneth structure.

In the other direction, if we have a Künneth structure $(\omega, \mathcal{F}, \mathcal{G})$, then by Theorem 6.10 the Levi–Civita connection ∇ of the associated neutral metric g is the Künneth connection. It therefore preserves the tangent bundles to the two Lagrangian foliations. It then follows from the definition of I as $\mathrm{Id}_{T\mathcal{F}} \ominus \mathrm{Id}_{T\mathcal{G}}$ that I commutes with ∇, and so (I, g) is a para-Kähler structure.   □

This shows that the notions of Künneth and para-Kähler structures are indeed equivalent.

Notes for Chapter 6

1. In bi-Lagrangian terminology, the canonical connection of an almost Künneth structure was probably first introduced by Hess [Hes-80, Hes-81]. In the integrable case it was discussed by Kanai [Kan-88], and it is often called the Kanai connection, especially in the dynamical systems literature. The integrable case also appears in Boyom [Boy-89], which we found very useful for our presentation, in Tabachnikov [Tab-93], and in many other references.
2. To the best of our knowledge, Theorem 6.10 was first made explicit by Etayo Gordejuela and Santamaría [ES-01, Theorem 3].
3. The equivalence between para-Kähler and bi-Lagrangian structures is well known. It appears for example in [CFG-96, ES-01]. See also [EST-06].

7

The Curvature of a Künneth Structure

In this chapter we discuss the curvature of the Künneth connection. First we work out some general properties of the curvature tensor, then we prove a theorem showing that the curvature is the precise obstruction for the validity of the simplest possible Darboux Theorem for Künneth structures. We then present some examples of vanishing and non-vanishing curvature, and we work out the Ricci and scalar curvatures of the associated pseudo Riemannian metric. This leads naturally to a discussion of the Einstein condition in this setting.

In the final section of this chapter we consider Künneth structures compatible with a positive-definite Kähler metric, and we show that in this case the Künneth structure and the Kähler metric are flat.

7.1 Symmetries of the Künneth Curvature Tensor

Let (ω, F, G) be an almost Künneth structure on M, and ∇ its Künneth connection. In this section we examine the curvature

$$R(X, Y)Z = \nabla_X \nabla_Y Z - \nabla_Y \nabla_X Z - \nabla_{[X,Y]} Z \qquad (7.1)$$

of ∇ and relate it to the Darboux Theorem for Lagrangian (bi-)foliations.

In this discussion we will use the following lemma.

Lemma 7.1 *The identity*

$$D([X, Y], Z) = D(X, D(Y, Z)) - D(Y, D(X, Z))$$

holds for all vector fields X, Y and Z on M.

85

Proof   By Definition 6.2, which defines the operator D, we have

$$
\begin{aligned}
i_{(D(X,D(Y,Z))-D(Y,D(X,Z)))}\omega &= L_X i_{D(Y,Z)}\omega - L_Y i_{D(X,Z)}\omega \\
&= L_X L_Y i_Z \omega - L_Y L_X i_Z \omega \\
&= L_{[X,Y]} i_Z \omega \\
&= i_{D([X,Y],Z)}\omega \, .
\end{aligned}
$$

The claim now follows from the non-degeneracy of ω.   □

The curvature tensor of the Levi–Civita connection of a Riemannian metric g has the following well-known symmetry:

$$
g(R(X, Y)Z, T) = -g(Z, R(X, Y)T) \, .
$$

We adapt this to the almost Künneth situation at hand.

Lemma 7.2   *Let ω be a non-degenerate two-form on M, and ∇ an affine connection compatible with ω. Then the curvature tensor R of ∇ has the following symmetry:*

$$
\omega(R(X, Y)Z, T) = \omega(R(X, Y)T, Z)
$$

for all vector fields X, Y, Z and T on M.

This holds in particular for the Künneth connection of any almost Künneth structure.

Proof   *Mutatis mutandis* we argue as in the Riemannian setting. Compatibility between ∇ and ω gives us

$$
\begin{aligned}
\omega(\nabla_X \nabla_Y Z, T) &= L_X(\omega(\nabla_Y Z, T)) - \omega(\nabla_Y Z, \nabla_X T) \\
&= L_X(\omega(\nabla_Y Z, T)) - L_Y(\omega(Z, \nabla_X T)) + \omega(Z, \nabla_Y \nabla_X T) \, .
\end{aligned}
$$

Interchanging the rôles of X and Y and subtracting the result from the above identity we obtain

$$
\begin{aligned}
\omega(\nabla_X \nabla_Y Z - \nabla_Y \nabla_X Z, T) &= L_X(\omega(\nabla_Y Z, T) + \omega(Z, \nabla_Y T)) \\
&\quad - L_Y(\omega(Z, \nabla_X T) + \omega(\nabla_X Z, T)) \\
&\quad - \omega(Z, \nabla_X \nabla_Y T - \nabla_Y \nabla_X T) \\
&= L_X L_Y(\omega(Z, T)) - L_Y L_X(\omega(Z, T)) \\
&\quad - \omega(Z, \nabla_X \nabla_Y T - \nabla_Y \nabla_X T) \\
&= L_{[X,Y]}(\omega(Z, T)) - \omega(Z, \nabla_X \nabla_Y T - \nabla_Y \nabla_X T) \, .
\end{aligned}
$$

Compatibility between ∇ and ω also gives us

$$
L_{[X,Y]}(\omega(Z, T)) = \omega(\nabla_{[X,Y]} Z, T) + \omega(Z, \nabla_{[X,Y]} T) \, .
$$

Together the two calculations imply

$$\omega(R(X,Y)Z,T) = -\omega(Z,R(X,Y)T)$$

and hence the claim.   □

To obtain further symmetries for the curvature tensor of the Künneth connection, we will assume at least partial integrability. If one of the Lagrangian subbundles, say F, is integrable to a foliation $\mathcal{F}$, then according to Remark 6.11 the Künneth connection on $T\mathcal{F}$ along the leaves of $\mathcal{F}$ is the Bott connection, and according to Proposition 4.9 that connection is flat. The following proposition shows that more is true, in that the Künneth connection on the Lagrangian complement G is also flat. This vanishing of curvature along the leaves of the foliations making up a Künneth structure will lead to additional symmetries of the remaining components of the curvature tensor.

Proposition 7.3 *Let (ω, F, G) be an almost Künneth structure. If F is integrable to a foliation $\mathcal{F}$, then the Künneth connection ∇ on TM is flat along every leaf of $\mathcal{F}$.*

Proof Let X, Y be sections of $F = T\mathcal{F}$. If Z is also a section of F, then

$$R(X,Y)Z = \nabla_X\nabla_Y Z - \nabla_Y\nabla_X Z - \nabla_{[X,Y]}Z$$
$$= D(X, D(Y,Z)) - D(Y, D(X,Z)) - D([X,Y],Z)$$

vanishes by Lemma 7.1. Therefore F is flat when restricted to a leaf of $\mathcal{F}$. (As explained above, this also follows from Proposition 4.9.)

If Z is a section of G, then by the Jacobi identity

$$R(X,Y)Z = [X,[Y,Z]_G]_G - [Y,[X,Z]_G]_G - [[X,Y],Z]_G$$
$$= ([X,[Y,Z]] + [Y,[Z,X]] + [Z,[X,Y]])_G = 0 .$$

In the step from the first to the second line we have used the fact that the terms

$$[X,[Y,Z]_F]_G \quad \text{and} \quad [Y,[X,Z]_F]_G$$

vanish by the integrability of F. Therefore G is also flat when restricted to a leaf of $\mathcal{F}$.   □

Proposition 7.3 shows that many components of the curvature tensor of ∇ vanish if both F and G are integrable. The only interesting components are the *mixed curvatures*, i.e. $R(X,Y)Z$ where X and Y are tangent to different foliations.

In the integrable case we have the following collection of symmetries for the curvature tensor.

Theorem 7.4   *The curvature tensor of the Künneth connection of a Künneth structure $(\omega, \mathcal{F}, \mathcal{G})$ satisfies the following identities.*

(i) *For arbitrary vectors X, Y, Z and $T \in TM$ we have*

$$R(X, Y)Z + R(Y, Z)X + R(Z, X)Y = 0 \qquad (7.2)$$

and

$$\omega(R(X, Y)Z, T) = \omega(R(X, Y)T, Z) . \qquad (7.3)$$

(ii) *For an arbitrary vector $X \in TM$ and vectors Y and Z both tangent either to $\mathcal{F}$ or to $\mathcal{G}$ we have*

$$R(X, Y)Z = R(X, Z)Y . \qquad (7.4)$$

(iii) *For vectors X and Z both tangent to $\mathcal{F}$ or to $\mathcal{G}$, and Y and T both tangent to $\mathcal{F}$ or to $\mathcal{G}$, we have*

$$\omega(R(X, Y)Z, T) = \omega(R(Z, T)X, Y) . \qquad (7.5)$$

Proof   Since the Künneth connection is torsion free in the integrable case, the identity (7.2) holds for its curvature. It is just the Bianchi identity, known to hold for any torsion-free affine connection. Identity (7.3) was proved in a more general setting in Lemma 7.2.

Identity (7.4) follows immediately from the Bianchi identity (7.2) and the skew-symmetry of $R(X, Y)Z$ in X and Y using Proposition 7.3.

We derive the identity (7.5) as follows:

$$
\begin{aligned}
\omega(R(X, Y)Z, T) &= -\omega(R(Y, X)Z, T) && \text{because } R(X, Y) = -R(Y, X)\\
&= -\omega(R(Y, Z)X, T) && \text{by (7.4)}\\
&= \omega(R(Z, Y)X, T) && \text{because } R(Y, Z) = -R(Z, Y)\\
&= \omega(R(Z, Y)T, X) && \text{by (7.3)}\\
&= \omega(R(Z, T)Y, X) && \text{by (7.4)}\\
&= \omega(R(Z, T)X, Y) && \text{by (7.3)} .
\end{aligned}
$$

This completes the proof of the theorem.   □

The trivial example of a product Künneth structure is of course flat.

Example 7.5   Consider the standard linear Künneth structure on $\mathbb{R}^{2n} = \mathbb{R}^n \times (\mathbb{R}^n)^*$ with the standard symplectic form ω_0 and the affine subspaces parallel to the two factors being the leaves of the Lagrangian foliations $\mathcal{F}_0$ and $\mathcal{G}_0$. All these structures are parallel for the standard flat connection ∇_0. As this connection is torsion free, it is the Künneth connection.

We now prove that any flat Künneth structure looks locally like this example; in other words, the Darboux Theorem in its simplest form holds for flat Künneth structures. Obviously it cannot hold for a non-flat structure, since the Künneth connection is intrinsically and uniquely defined by the Künneth structure.

Theorem 7.6 *The curvature of the Künneth connection of a Künneth structure $(\omega, \mathcal{F}, \mathcal{G})$ vanishes if and only if the Künneth structure is locally isomorphic to the standard structure $(\omega_0, \mathcal{F}_0, \mathcal{G}_0)$ on $\mathbb{R}^{2n}$.*

Proof We have to prove only that a flat Künneth structure is locally standard. In the proof we use that a flat connection determines a well-defined parallel transport in a simply connected neighbourhood of a point, which is independent of any choices of paths.

Let $p \in M$. It is clear from Proposition 2.16 that the linear Künneth structure on the tangent space at p given by $T_p\mathcal{F}$, $T_p\mathcal{G}$ and ω_p has the standard form in a suitable basis, i.e. there exists a basis $f_1, \ldots, f_n \in T_p\mathcal{F}$, $g_1, \ldots, g_n \in T_p\mathcal{G}$ of T_pM such that

$$\omega(f_i, f_j) = 0 = \omega(g_i, g_j) ,$$
$$\omega(f_i, g_j) = \delta_{ij} .$$

We want to extend this basis to a whole neighbourhood U, diffeomorphic to an open ball in $\mathbb{R}^{2n}$, of the point p.

First, note that the Künneth connection restricts, by its definition, to a connection on $T\mathcal{F} \to M$, which is again flat. Therefore, because U is simply connected, the parallel transport of vectors in $T\mathcal{F}$ along curves in this neighbourhood of p is independent of the curve chosen.

In this way we get extensions of $f_1, \ldots, f_n$ to vector fields $F_1, \ldots, F_n \in \Gamma(T\mathcal{F})$ on U that are parallel with respect to the Künneth connection. Similarly, we get parallel vector fields $G_1, \ldots, G_n \in \Gamma(T\mathcal{G})$ that restrict to the $g_1, \ldots, g_n$ at p.

Now we use the fact that the Künneth connection is compatible with ω to see that the equations

$$\omega(F_i, F_j) = 0 = \omega(G_i, G_j) ,$$
$$\omega(F_i, G_j) = \delta_{ij} \tag{7.6}$$

still hold everywhere. Finally, because the connection is torsion free, and the F_i and G_i are parallel, the commutators between all these basis vector fields vanish. This implies that there exist local coordinates $(x_1, \ldots, x_n, y_1, \ldots, y_n)$ around p such that $F_i = \partial_{x_i}$ and $G_i = \partial_{y_i}$. These coordinates establish the local isomorphism of the Künneth structure on M with the standard Künneth structure on $\mathbb{R}^{2n}$, since (7.6) shows that ω takes the standard form ω_0 in these coordinates, and at the same time the x_i and the y_j are coordinates along the leaves of $\mathcal{F}$ and $\mathcal{G}$ respectively.   □

Remark 7.7   For a general Künneth structure we can always find, according to Proposition 4.14, a chart such that ω and one of the Lagrangian foliations are standard. Theorem 7.6 shows that we can do this simultaneously for both foliations if and only if the Künneth connection is flat.

In general, even when a Künneth structure on a compact manifold M is flat, we do not have much control over the global geometry and topology of M, since usually one knows nothing about completeness of the Künneth connection. The Künneth connection is the Levi–Civita connection of an indefinite pseudo-Riemannian metric, and indefiniteness means that compactness does not imply completeness. Therefore, the standard local models provided by Theorem 7.6 do not lead to a standard global model on the universal covering of M. In dimension 2 one can say more, because in this case the manifold M is a torus and the neutral metric is Lorentzian. It is a result of Carrière [Car-94] that flat Lorentzian tori are complete. Therefore, their universal coverings are isometric to the Minkowski plane. This allows one to prove the following characterisation of Künneth structures with non-compact automorphism groups.

Theorem 7.8   *Any Künneth structure on T^2 with a non-compact automorphism group is induced by a linear Anosov isometry of the Minkowski plane.*

In other words, Corollary 5.19 accounts for all Künneth structures with non-compact automorphism group on T^2.

We only outline the proof of Theorem 7.8. First of all, any Lorentz metric on T^2 with non-compact isometry group must be flat by a result of Mounoud [Mou-03]. It is complete [Car-94], and so the universal covering is isometric to the Minkowski plane. Lifting isometries to the universal covering, one sees that the non-compactness of the isometry group downstairs means that there is an isometry which lifts to a hyperbolic element of $O(1, 1)$. This is an Anosov automorphism which induces the Künneth structure we started with.

7.2 Concrete Curvature Computations

7.2.1 Interesting Flat Examples

We want to show that there exist closed manifolds that are neither tori nor covered by tori, yet have a flat Künneth structure.

First consider a linear Anosov symplectomorphism φ of T^2 as in Example 5.20. In this case the symplectic form we consider is the standard one, $\omega_0 = dx \wedge dy$, and the two Lagrangian foliations are linear foliations spanned by constant linear combinations of the basis vectors ∂_x and ∂_y. The standard flat connection ∇_0, for which ∂_x and ∂_y are parallel, preserves the foliations and is compatible with ω_0. Therefore it is the Künneth connection.

Next we form the symplectic mapping torus $N = M(\varphi) \times S^1$ of φ and suspend the Künneth structure. The solvmanifold N is a T^2-bundle over T^2 and by Theorem 5.23 the resulting Künneth structure on it is locally the product of the above flat Künneth structure on the fibre, and whichever Künneth structure we want to use on the base. By Example 6.7 the Künneth connection is locally the direct sum connection of the Künneth connections on the base and the fibre. Therefore, if we use a flat Künneth structure on the base, then the suspended Künneth structure on N is also flat.

7.2.2 Easy Non-flat Examples

Already on $\mathbb{R}^2$ and T^2 there are Künneth structures with non-vanishing curvature. We consider a concrete example. Let x, y be linear coordinates on $\mathbb{R}^2$ with coordinate vector fields ∂_x, ∂_y. The two-form

$$\omega = (2 + \sin(2\pi x)\sin(2\pi y))dx \wedge dy = f(x, y)dx \wedge dy \qquad (7.7)$$

is nowhere zero and hence symplectic. The complementary foliations $\mathcal{F}$ and $\mathcal{G}$ spanned by the coordinate vector fields are Lagrangian. This defines a Künneth structure on $\mathbb{R}^2$.

We want to calculate the vector field $D(\partial_x, \partial_x)$. If X and Z are commuting vector fields then we know from the definition of D that

$$\omega(D(X, Y), Z) = L_X(\omega(Y, Z)).$$

For the basis vectors ∂_x, ∂_y this implies:

$$\omega(D(\partial_x, \partial_x), \partial_x) = 0,$$
$$\omega(D(\partial_x, \partial_x), \partial_y) = \partial_x \omega(\partial_x, \partial_y) = 2\pi \cos(2\pi x)\sin(2\pi y),$$

and hence

$$D(\partial_x, \partial_x) = \frac{2\pi}{f}\cos(2\pi x)\sin(2\pi y)\partial_x.$$

By the definition of the Künneth connection we obtain

$$\nabla_{\partial_x}\partial_x = D(\partial_x, \partial_x) = \frac{2\pi}{f}\cos(2\pi x)\sin(2\pi y)\partial_x \,,$$

$$\nabla_{\partial_y}\partial_x = [\partial_y, \partial_x]_F = 0 \,.$$

Now we calculate a mixed curvature term:

$$
\begin{aligned}
R(\partial_x, \partial_y)\partial_x &= \nabla_{\partial_x}\nabla_{\partial_y}\partial_x - \nabla_{\partial_y}\nabla_{\partial_x}\partial_x - \nabla_{[\partial_x,\partial_y]}\partial_x \\
&= -\nabla_{\partial_y}\nabla_{\partial_x}\partial_x \\
&= -\partial_y\left(\frac{2\pi}{f}\cos(2\pi x)\sin(2\pi y)\right)\partial_x \\
&= -\left(\frac{4\pi^2}{f}\cos(2\pi x)\cos(2\pi y)\right. \\
&\qquad \left. -\frac{4\pi^2}{f^2}\cos(2\pi x)\sin(2\pi y)\sin(2\pi x)\cos(2\pi y)\right)\partial_x \\
&= -\frac{8\pi^2}{f^2}\cos(2\pi x)\cos(2\pi y)\partial_x \,.
\end{aligned}
$$

For example, in the point $(x, y) = (0, 0)$ we have

$$(R(\partial_x, \partial_y)\partial_x)(0, 0) = -2\pi^2\partial_x \,,$$

and so the curvature certainly does not vanish identically.

Note that the Künneth structure is invariant under the standard action of $\mathbb{Z}^2$ on $\mathbb{R}^2$ by addition, hence we get a Künneth structure with non-vanishing curvature on T^2 (the curvature vanishes on T^2 only along the pairs of opposite circles corresponding to $x = \pm\frac{1}{4}$ and $y = \pm\frac{1}{4}$). Taking products we get Künneth structures on higher-dimensional tori with non-zero curvature.

Moreover, if we rescale the symplectic form ω by $\frac{1}{2}$ and modify the second summand in equation (7.7) with a suitable cut-off function, we see that there exists a symplectic form on $\mathbb{R}^{2n}$ that is the standard form outside of a small neighbourhood of the origin, such that the Künneth structure with the standard foliations has curvature not identically zero.

Let M be any flat Künneth manifold. If we take out a small Darboux neighbourhood of a point as in Theorem 7.6 and glue in a copy of the Künneth structure discussed above, then we get a Künneth structure with non-vanishing curvature. Therefore we can always deform a flat Künneth structure so that it becomes curved.

7.3 The Ricci and Scalar Curvatures

Let $(\omega, \mathcal{F}, \mathcal{G})$ be a Künneth structure on M, and g the associated metric of neutral signature. In this section we want to study the Ricci curvature of g. One could do this more generally for almost Künneth structures, but we focus on the integrable case because only in this case does the Levi–Civita connection of g agree with the Künneth connection, so that the curvature of g is the curvature of the Künneth structure.

We first recall the definition of the Ricci tensor and its most basic property.

Definition 7.9 The *Ricci curvature* $\mathrm{Ric}(X, Y)$ of vectors $X, Y \in TM$ is defined as the trace of the map

$$TM \longrightarrow TM$$
$$Z \longmapsto R(Z, X)Y \,.$$

Lemma 7.10 *The Ricci tensor is symmetric,*

$$\mathrm{Ric}(X, Y) = \mathrm{Ric}(Y, X) \quad \forall X, Y \in TM \,.$$

Proof This result holds for the Ricci curvature of the Levi–Civita connection of any pseudo-Riemannian metric g, because in this situation the curvature satisfies

$$g(R(X, Y)Z, T) = -g(R(X, Y)T, Z)$$

and the Bianchi identity

$$R(X, Y)Z + R(Y, Z)X + R(Z, X)Y = 0 \,.$$

The first identity implies for any vector $e_i \in T_pM$

$$g(R(X, Y)e_i, e_i) = 0 \,,$$

and then the Bianchi identity implies

$$g(R(e_i, X)Y, e_i) = -g(R(Y, e_i)X, e_i)$$
$$= g(R(e_i, Y)X, e_i) \,.$$

Choosing the e_i as an orthonormal basis (in the pseudo-Riemannian sense), it follows that the trace that defines the Ricci curvature is symmetric.   □

The following proposition gives a simple formula for the Ricci curvature of a Künneth structure. Let $p \in M$ be a point. We fix a symplectic basis $f_1, \ldots, f_n \in$

$T_p\mathcal{F}, g_1, \ldots, g_n \in T_p\mathcal{G}$ adapted to the foliations as before. Let $r_p \colon T_pM \longrightarrow T_pM$ be the linear map

$$r_p(X) = \sum_{i=1}^{n} R(f_i, g_i)X. \tag{7.8}$$

Proposition 7.11 *For any Künneth structure $(\omega, \mathcal{F}, \mathcal{G})$ the map r_p does not depend on the choice of symplectic basis f_i, g_j adapted to the Lagrangian foliations $\mathcal{F}$ and $\mathcal{G}$ as above. It defines an endomorphism of the tangent bundle*

$$r \colon TM \longrightarrow TM$$

preserving the splitting $TM = T\mathcal{F} \oplus T\mathcal{G}$. The Ricci curvature can be calculated in terms of r:

$$\mathrm{Ric}(X, Y) = \omega(r(X), Y) \quad \forall X, Y \in TM. \tag{7.9}$$

Proof It suffices to prove the formula for the Ricci curvature, because this implies that r is well defined independent of choices. The trace of an endomorphism is the sum over the diagonal entries of the matrix representation in any basis. We will use the g-orthonormal basis made up of the vectors

$$e_i^{\pm} = \frac{1}{\sqrt{2}}(f_i \pm g_i).$$

We have $g(e_i^{\pm}, e_i^{\pm}) = \pm 1$, so the formula for the trace becomes

$$\mathrm{Ric}(X, Y) = \sum_{i=1}^{n} g(R(e_i^+, X)Y, e_i^+) - \sum_{i=1}^{n} g(R(e_i^-, X)Y, e_i^-).$$

Using the definition of the $e_i^{\pm}$ and the relation between g and ω, this gives

$$\begin{aligned}
\mathrm{Ric}(X, Y) &= \frac{1}{2}\sum_i g(R(f_i + g_i, X)Y, f_i + g_i) - \frac{1}{2}\sum_i g(R(f_i - g_i, X)Y, f_i - g_i) \\
&= \frac{1}{2}\sum_i \left(\omega(R(f_i + g_i, X)Y, -f_i + g_i) + \omega(R(f_i - g_i, X)Y, f_i + g_i)\right) \\
&= \sum_i \left(\omega(R(f_i, X)Y, g_i) - \omega(R(g_i, X)Y, f_i)\right).
\end{aligned}$$

Now the formulas (7.3) and (7.2) from Theorem 7.4 allow us to conclude

$$
\begin{aligned}
\mathrm{Ric}(X, Y) &= \sum_i \left(\omega(R(f_i, X)Y, g_i) - \omega(R(g_i, X)Y, f_i) \right) \\
&= \sum_i \left(\omega(R(f_i, X)g_i, Y) - \omega(R(g_i, X)f_i, Y) \right) \\
&= \sum_i \left(\omega(R(f_i, X)g_i, Y) + \omega(R(X, g_i)f_i, Y) \right) \\
&= \sum_{i=1}^n \omega(R(f_i, g_i)X, Y) \,.
\end{aligned}
$$

This completes the proof. □

The following consequence of formula (7.9) shows that only the mixed terms of the Ricci tensor can be non-zero, just as for the curvature tensor.

Corollary 7.12 *The Ricci curvature* $\mathrm{Ric}(X, Y)$ *vanishes if X and Y are both tangent to $\mathcal{F}$ or tangent to $\mathcal{G}$, i.e.*

$$
\mathrm{Ric}\,|_{T\mathcal{F}} \equiv 0 \equiv \mathrm{Ric}\,|_{T\mathcal{G}} \,.
$$

Proof This follows from (7.9) since r preserves the subbundles $T\mathcal{F}$ and $T\mathcal{G}$ and both are Lagrangian. □

The above discussion leads quite directly to a simple formula for the scalar curvature of a Künneth structure. In general, the scalar curvature of a pseudo-Riemannian metric is defined as follows.

Definition 7.13 Let g be a pseudo-Riemannian metric on a manifold M. Then there is a unique smooth section $A \in \Gamma(\mathrm{End}(TM))$ for which

$$
\mathrm{Ric}(X, Y) = g(AX, Y) \,.
$$

The *scalar curvature* of g is defined as the trace of A:

$$
\mathrm{scal} = \mathrm{tr}(A) \,.
$$

Choosing a basis $e_1, \ldots, e_n$ of T_pM of pairwise g-orthogonal vectors so that $g(e_i, e_i) = \pm 1$ for all $i = 1, \ldots, n$, the value of the scalar curvature at p can be computed by the formula

$$
\mathrm{scal}(p) = \sum_{i=1}^n \mathrm{Ric}(e_i, e_i)g(e_i, e_i) \,.
$$

Again we fix a symplectic basis $f_1, \ldots, f_n \in T_p\mathcal{F}$, $g_1, \ldots, g_n \in T_p\mathcal{G}$ as above.

Proposition 7.14   *The scalar curvature of a Künneth structure is given by the following equivalent expressions:*

$$\text{scal} = 2 \sum_{j=1}^{n} \text{Ric}(f_j, g_j) = 2 \cdot \text{tr}(r|_{T\mathcal{F}}) . \tag{7.10}$$

Proof   In the associated orthonormal basis $e_i^{\pm}$ the scalar curvature is

$$\text{scal} = \sum_{j=1}^{n} \text{Ric}(e_j^+, e_j^+) - \sum_{j=1}^{n} \text{Ric}(e_j^-, e_j^-) .$$

This implies

$$\begin{aligned}
\text{scal} &= \frac{1}{2} \sum_{j=1}^{n} \text{Ric}(f_j + g_j, f_j + g_j) - \frac{1}{2} \sum_{j=1}^{n} \text{Ric}(f_j - g_j, f_j - g_j) \\
&= \sum_{j=1}^{n} \text{Ric}(f_j, g_j) + \sum_{j=1}^{n} \text{Ric}(f_j, g_j) \\
&= 2 \sum_{j=1}^{n} \text{Ric}(f_j, g_j) ,
\end{aligned}$$

where we used the fact that the Ricci curvature is symmetric and vanishes if both arguments are in $T\mathcal{F}$ or in $T\mathcal{G}$. This proves the first equality. To show the second one we note that

$$\text{tr}(r|_{T\mathcal{F}}) = \sum_{j=1}^{n} \omega(r(f_j), g_j) = \sum_{i=1}^{n} \text{Ric}(f_j, g_j) .$$

This completes the proof.   □

Remark 7.15   Recall from the discussion in Section 5.2 of Chapter 5 that a given (almost) Künneth structure determines the induced metric g only up to sign. If we reverse the sign of a pseudo-Riemannian metric, then the Levi–Civita connection stays the same because $-g$ is parallel for ∇ and ∇ is torsion free. This implies that the Riemann curvature tensor and the Ricci tensor also remain the same, but the scalar curvature changes its sign.

7.4 Künneth–Einstein Structures

We now investigate the Einstein condition for the neutral metric g associated with a Künneth structure $(\omega, \mathcal{F}, \mathcal{G})$.

Definition 7.16   The Künneth structure $(\omega, \mathcal{F}, \mathcal{G})$ is called *Künneth–Einstein* if the associated neutral metric g is Einstein, i.e. Ric $= \lambda g$ for some constant $\lambda \in \mathbb{R}$.

If we replace g by $-g$, then according to Remark 7.15 the Ricci tensor stays the same and the scalar curvature changes sign. It follows that the Einstein constant λ of a Künneth–Einstein structure changes sign if we exchange the rôles of $\mathcal{F}$ and $\mathcal{G}$, or replace the almost product structure I by $-I$. We will usually fix the standard choice for I as in Section 5.2 of Chapter 5 and exchange $\mathcal{F}$ and $\mathcal{G}$ to change the sign of the Einstein constant.

Note that both Ric and g vanish identically along the foliations $\mathcal{F}$ and $\mathcal{G}$. Proposition 7.11 leads to the following characterisation of the Einstein condition in this setting.

Proposition 7.17   *A Künneth structure is Künneth–Einstein with constant $\lambda \in \mathbb{R}$ if and only if*

$$r \equiv \lambda I : TM \longrightarrow TM ,$$

where $I : TM \longrightarrow TM$ is the almost product structure $\mathrm{Id}_{T\mathcal{F}} \ominus \mathrm{Id}_{T\mathcal{G}}$ defined by the Künneth structure. Equivalently,

$$r|_{T\mathcal{F}} \equiv \lambda \, \mathrm{Id}_{T\mathcal{F}} .$$

Proof   The identity (7.9) shows that Ric $= \lambda g$ is equivalent to

$$\omega\left(r(X), Y\right) = \lambda g(X, Y) = \omega(\lambda I X, Y) .$$

The non-degeneracy of ω now proves the first claim.

For the second claim we use the symmetry of the Ricci tensor, Ric$(X, Y) =$ Ric(Y, X), and apply formula (7.9) to both sides. Since the mixed Ricci curvatures vanish, we may assume $X \in T\mathcal{F}$ and $Y \in T\mathcal{G}$. We see that $r = \lambda \mathrm{Id}$ on $T\mathcal{F}$ is equivalent to $r = -\lambda \mathrm{Id}$ on $T\mathcal{G}$, again by the non-degeneracy of ω. This completes the proof. $\square$

Example 7.18   The simplest examples of Künneth–Einstein structures are the flat ones, since they are in particular Ricci-flat. Thus the standard Künneth structure on $\mathbb{R}^{2n}$ and on T^{2n} is Künneth–Einstein with constant $\lambda = 0$.

We now discuss the two-dimensional case in some detail. Let $(M^2, \omega, \mathcal{F}, \mathcal{G})$ be a surface with a Künneth structure. The metric g then has signature $(1, 1)$ and is Lorentzian. At a point $p \in M$ we choose a symplectic basis (f, g) for $T_p M$ adapted to the foliations and consider the orthonormal basis

$$e^+ = \frac{1}{\sqrt{2}}(f + g) , \quad e^- = \frac{1}{\sqrt{2}}(f - g) .$$

The only non-trivial component of the Riemann curvature tensor is given by

$$g(R(f,g)f,g) = \omega(R(f,g)f,g) = \text{Ric}(f,g) \, .$$

If the structure is Künneth–Einstein with $\text{Ric} = \lambda g$, then

$$g(R(f,g)f,g) = \lambda \, .$$

We first note the following in the case of closed surfaces.

Theorem 7.19 *Any Künneth–Einstein structure on a closed surface M is flat.*

Proof The only closed oriented surface that admits a Lorentz metric is the torus $M = T^2$, hence $\chi(M) = 0$. According to the Gauss–Bonnet–Chern Theorem [Ave-62, BN-84] for Lorentz metrics we have

$$\chi(M) = \frac{1}{\pi} \int_M \lambda \, \text{dvol}_g = \frac{1}{\pi} \lambda \text{vol}_g(M) \, .$$

This implies that $\lambda = 0$. Like in the Riemannian case, the vanishing of the Gauss curvature implies that the Lorentz metric is flat. □

We will extend Theorem 7.19 to closed four-manifolds in Theorem 10.24. There the conclusion will be that every Künneth–Einstein structure on a closed four-manifold is Ricci-flat. Unlike in dimension 2, Ricci-flatness does not imply flatness in dimension 4 and higher.

For open surfaces Theorem 7.19 is no longer true, as the following simple examples show. We will see that there is a Künneth–Einstein structure on $\mathbb{R}^2$ whose induced Lorentz metric is the de Sitter or anti-de Sitter metric, depending on the sign of the Einstein constant.

Continuing the discussion in Subsection 7.2.2 we look for Künneth–Einstein structures that are conformal rescalings of the standard Künneth structure:

$$\omega = e^A dx \wedge dy \, ,$$
$$\mathcal{F} = \text{span}\{\partial_x\} \, ,$$
$$\mathcal{G} = \text{span}\{\partial_y\} \, ,$$

where A is a smooth function on (a subset of) $\mathbb{R}^2$. It is clear that ω is symplectic, hence $(\omega, \mathcal{F}, \mathcal{G})$ form a Künneth structure. In fact, ω must have this form (up to sign) if the Lagrangian foliations of the Künneth structure on $\mathbb{R}^2$ are the trivial ones. We rescale

$$f = e^{-A/2}\partial_x \, ,$$
$$g = e^{-A/2}\partial_y \, ,$$

so that $\omega(f, g) = 1$; in other words f and g form a symplectic basis. According to Proposition 7.17 the associated metric is Einstein if and only if

$$R(f, g)f = \lambda f ,$$

where λ is a constant. Equivalently

$$R(\partial_x, \partial_y)\partial_x = \lambda e^A \partial_x .$$

We have

$$R(\partial_x, \partial_y)\partial_x = \nabla_{\partial_x}\nabla_{\partial_y}\partial_x - \nabla_{\partial_y}\nabla_{\partial_x}\partial_x - \nabla_{[\partial_x,\partial_y]}\partial_x .$$

By the definition of the Künneth connection

$$\nabla_{\partial_y}\partial_x = [\partial_y, \partial_x]_F = 0 ,$$
$$\nabla_{\partial_x}\partial_x = D(\partial_x, \partial_x) .$$

Therefore

$$R(\partial_x, \partial_y)\partial_x = -[\partial_y, D(\partial_x, \partial_x)]_F .$$

We have

$$\omega(D(\partial_x, \partial_x), \partial_y) = \partial_x(\omega(\partial_x, \partial_y))$$
$$= \partial_x e^A$$
$$= e^A(\partial_x A) .$$

Hence

$$D(\partial_x, \partial_x) = (\partial_x A)\partial_x$$

and

$$R(\partial_x, \partial_y)\partial_x = -(\partial_y\partial_x A)\partial_x .$$

We have proved the following.

Proposition 7.20 *The Künneth structure*

$$\omega = e^A dx \wedge dy ,$$
$$\mathcal{F} = \mathrm{span}\{\partial_x\} ,$$
$$\mathcal{G} = \mathrm{span}\{\partial_y\}$$

on $\mathbb{R}^2$ is Einstein with constant λ if and only if A satisfies the equation

$$\partial_y\partial_x A + \lambda e^A = 0 . \tag{7.11}$$

This equation is a Liouville type equation for Lorentz metrics of signature $(1, 1)$. Here is a concrete solution of equation (7.11).

Proposition 7.21   *For a real constant $\lambda > 0$ consider the symplectic form on $\mathbb{R}^2$ given by*

$$\omega = \frac{2}{\lambda(x-y)^2} dx \wedge dy .$$

Then the Künneth structure $(\omega, \mathcal{F}, \mathcal{G})$ with the foliations $\mathcal{F} = \mathrm{span}\{\partial_x\}$ and $\mathcal{G} = \mathrm{span}\{\partial_y\}$ is Einstein with constant $\lambda > 0$. Reversing the rôle of the two foliations makes the structure Einstein with constant $-\lambda < 0$.

The symplectic form, Lorentz metric and curvature are all singular along the line $x = y$.

Proof   We have

$$A = \ln\left(\frac{2}{\lambda(x-y)^2}\right) = \ln\left(\frac{2}{\lambda}\right) - 2\ln(x-y) .$$

Hence

$$\begin{aligned}
\partial_y\partial_x A &= \partial_y \frac{-2}{x-y} \\
&= \frac{-2}{(x-y)^2} .
\end{aligned}$$

Since

$$\lambda e^A = \frac{2}{(x-y)^2} ,$$

the claim follows.   □

Remark 7.22   We can get a non-singular Künneth structure on a half-plane diffeomorphic to $\mathbb{R}^2$ if we restrict to $x - y < 0$ or $x - y > 0$ (half-plane above or below the diagonal $x = y$).

The associated Lorentz metric is given by

$$g = \pm\frac{2}{\lambda(x-y)^2}(dx \otimes dy + dy \otimes dx) ,$$

where the sign $\pm$ corresponds to the case of positive and negative Einstein constant. Defining so-called Poincaré coordinates

$$t = x + y , \quad z = x - y ,$$

the metric takes the form

$$g = \pm\frac{1}{\lambda z^2}(dt \otimes dt - dz \otimes dz) .$$

This shows that the metric g is the so-called de Sitter metric (positive Einstein constant) or anti-de Sitter metric (negative Einstein constant) on two-dimensional spacetime. These metrics are the Lorentz analogues of spherical and hyperbolic metrics in Riemannian geometry.

Since it is well known that the de Sitter and anti-de Sitter metrics are Einstein (they are solutions to the Einstein field equations in a vacuum with non-zero cosmological constant), we could have started with these metrics on $\mathbb{R}^2$ and then looked for adapted Künneth structures.

Using suitable coordinates it is possible to extend these Lorentz metrics and hence the Künneth structures to non-singular Künneth–Einstein structures on the cylinder $S^1 \times \mathbb{R}$. We can also take the product of n copies of these Künneth structures on half-planes in $\mathbb{R}^2$ with the same Einstein constant to get Künneth–Einstein structures on open subsets of $\mathbb{R}^{2n}$ with non-zero Einstein constants (and similarly on $T^n \times \mathbb{R}^n$).

7.5 Kähler–Künneth Structures

It is natural to wonder whether Künneth, i.e. para-Kähler, structures can exist on Kähler manifolds in the usual sense, and be compatible with the Kähler structure. We now investigate this situation, imposing the most stringent compatibility conditions possible between the two structures. It will turn out that in this case the curvature vanishes identically, and so the manifolds that such a structure can exist on are limited to the flat Riemannian manifolds.

Suppose that the manifold M is endowed with a Kähler structure (J, g^K), meaning that J is an integrable complex structure and g^K is a positive-definite Riemannian metric such that

$$g^K(JX, JY) = g^K(X, Y)$$

and the two-form ω defined by

$$\omega(X, Y) = g^K(JX, Y)$$

is closed, and hence symplectic. The Levi–Civita connection ∇ of the Kähler metric g^K then preserves g^K and commutes with J, and hence is compatible with ω.

Proposition 7.23 *Suppose $\mathcal{F}$ is a Lagrangian foliation on a Kähler manifold M, and the Levi–Civita connection preserves $T\mathcal{F}$. Then $J(T\mathcal{F})$ is a parallel Lagrangian complement to $T\mathcal{F}$ and is integrable to a Lagrangian foliation $\mathcal{G}$. Hence $(\omega, \mathcal{F}, \mathcal{G})$ is a Künneth structure.*

Proof We first show that $J(T\mathcal{F})$ is complementary to $T\mathcal{F}$ and Lagrangian. First, the subbundle $J(T\mathcal{F})$ has the same rank as $T\mathcal{F}$. Second, for $X \in T\mathcal{F}$, we have

$$g^K(JX, Y) = \omega(X, Y) = 0 \quad \forall Y \in T\mathcal{F}$$

and hence $J(T\mathcal{F}) \subset (T\mathcal{F})^\perp$, where $(T\mathcal{F})^\perp$ denotes the g^K-orthogonal complement of $T\mathcal{F}$. Since g^K is positive definite and $T\mathcal{F}$ has rank half that of TM, this shows that $J(T\mathcal{F}) = (T\mathcal{F})^\perp$ and $J(T\mathcal{F})$ is complementary to $T\mathcal{F}$. It also follows that $J(T\mathcal{F})$ is Lagrangian, because

$$\omega(JX, JY) = \omega(X, Y) = 0$$

for all $X, Y \in T\mathcal{F}$.

Since $T\mathcal{F}$ is parallel for the Levi–Civita connection, which commutes with J, it follows that $J(T\mathcal{F})$ is also parallel, and hence integrable by Proposition 4.19. Hence $J(T\mathcal{F})$ integrates to a foliation $\mathcal{G}$ such that $(\omega, \mathcal{F}, \mathcal{G})$ is a Künneth structure. □

We encode this situation in the following definition.

Definition 7.24   A Künneth structure defined by a pair of orthogonal parallel Lagrangian foliations on a Kähler manifold will be called a *Kähler–Künneth structure*.

As one might expect, the Künneth connection of a Kähler–Künneth structure is just the Levi–Civita connection of the Kähler metric.

Proposition 7.25   *Let $(J, g^K, \omega, \mathcal{F}, \mathcal{G})$ be a Kähler–Künneth structure. Its Künneth connection is the Levi–Civita connection of the Kähler metric g^K.*

Proof By assumption the Levi–Civita connection preserves $T\mathcal{F}$ and $T\mathcal{G}$. Since it is compatible with ω and is torsion free, the claim follows from the uniqueness part of Corollary 6.9. □

We can reverse the above construction in the following way.

Proposition 7.26   *Let $(\omega, \mathcal{F}, \mathcal{G})$ be a Künneth structure on a manifold M. Suppose that J is an almost complex structure on M such that:*

(i) *J is compatible with ω, i.e. $g^K(X, Y) = \omega(X, JY)$ is a positive-definite Riemannian metric,*

(ii) *J commutes with the Künneth connection.*

Then (J, g^K) is a Kähler structure whose Levi–Civita connection is the Künneth connection of $(\omega, \mathcal{F}, \mathcal{G})$. Moreover, $\mathcal{F}$ and its g^K-orthogonal complement give rise to a Kähler–Künneth structure.

Proof Since the almost complex structure J commutes with the torsion-free Künneth connection, it is integrable, i.e. it defines a complex manifold structure on M.

Since the Künneth connection is compatible with ω and by assumption commutes with J, it follows that it preserves g^K. Since it is also torsion free, the Künneth connection is the Levi–Civita connection of g^K by the characterisation of the latter as the unique torsion-free connection compatible with g^K. This implies that (J, g^K) is a Kähler structure.

The subbundle $T\mathcal{F}$ is preserved by the Künneth connection, and since this equals the Levi–Civita connection of g^K, $T\mathcal{F}$ is parallel for the latter connection. We are therefore in the situation of Proposition 7.23. □

Note that we did not assume $J(T\mathcal{F}) = T\mathcal{G}$ in the proposition, and so these two Lagrangian complements of $T\mathcal{F}$ may be different in general.

The following theorem shows that Kähler–Künneth structures are rather special and occur only on flat Riemannian manifolds.

Theorem 7.27 *The curvature of a Kähler–Künneth structure vanishes identically.*

Proof Let $(J, g^K, \omega, \mathcal{F}, \mathcal{G})$ be a Kähler–Künneth structure on a smooth manifold M of dimension $2n$. Let $p \in M$ be a point and $f_1, \ldots, f_n$ a g^K-orthonormal basis for $T_p\mathcal{F}$. If we set $g_k = Jf_k$ then $g_1, \ldots, g_n$ are a g^K-orthonormal basis for $T_p\mathcal{G}$ and all $2n$ vectors together form an orthonormal basis for T_pM. We also write

$$e_1 = f_1, \ldots, e_n = f_n, \quad e_{n+1} = g_1, \ldots, e_{2n} = g_n$$

for this basis.

With respect to this basis, the curvature two-forms Ω_{ij} are defined by

$$R(X, Y)e_i = \sum_{j=1}^{2n} \Omega_{ij}(X, Y)e_j.$$

The curvature tensor of g^K is represented by the matrix $\Omega = (\Omega_{ij})$ of two-forms. To prove that g^K is flat we need to show that $\Omega \equiv 0$.

Recall from Section 5.1 of Chapter 5 that the splitting $TM = T\mathcal{F} \oplus T\mathcal{G}$ defines a bigrading on the differential forms on M.

Lemma 7.28 *The matrix Ω has the form*

$$\Omega = \begin{pmatrix} \Omega_{\mathcal{F}} & 0 \\ 0 & \Omega_{\mathcal{G}} \end{pmatrix},$$

where $\Omega_{\mathcal{F}}$ and $\Omega_{\mathcal{G}}$ are antisymmetric $(n \times n)$-matrices of two-forms. Furthermore, $\Omega_{\mathcal{F}} = \Omega_{\mathcal{G}}$. Each entry of these curvature matrices is a two-form of type $(1, 1)$ with respect to the bigrading induced by the splitting $TM = T\mathcal{F} \oplus T\mathcal{G}$.

Proof Since the matrix Ω represents the curvature of the Riemannian metric g^K with respect to an orthonormal basis, it is antisymmetric. Moreover, the splitting $TM = T\mathcal{F} \oplus T\mathcal{G}$ is parallel for the Levi–Civita connection, and so we must have

$$R(X, Y)(T\mathcal{F}) \subset T\mathcal{F} , \quad R(X, Y)(T\mathcal{G}) \subset T\mathcal{G} .$$

It follows that Ω has the form

$$\Omega = \begin{pmatrix} \Omega_{\mathcal{F}} & 0 \\ 0 & \Omega_{\mathcal{G}} \end{pmatrix} ,$$

where $\Omega_{\mathcal{F}}$ and $\Omega_{\mathcal{G}}$ are both antisymmetric. We have

$$\begin{aligned}
(\Omega_{\mathcal{G}})_{ij}(X, Y) &= g^K(R(X, Y)g_i, g_j) \\
&= g^K(R(X, Y)Jf_i, Jf_j) \\
&= g^K(JR(X, Y)f_i, Jf_j) \\
&= g^K(R(X, Y)f_i, f_j) \\
&= (\Omega_{\mathcal{F}})_{ij}(X, Y) ,
\end{aligned}$$

where we used the fact that J commutes with the Levi–Civita connection. This shows that $\Omega_{\mathcal{F}} = \Omega_{\mathcal{G}}$.

Finally, by Proposition 7.3 the Künneth connection is flat along the leaves of $\mathcal{F}$ and $\mathcal{G}$. This implies that all the curvature forms are of type $(1, 1)$. This completes the proof of the lemma. $\square$

It remains to determine $\Omega_{\mathcal{F}}$. To do this we first use the symmetry

$$R(X, Y)Z = R(X, Z)Y$$

whenever both Y and Z are tangent to $\mathcal{F}$. This was proved as identity (7.4) in Theorem 7.4. In the situation at hand it tells us that

$$(\Omega_{\mathcal{F}})_{ij}(X, f_k) = (\Omega_{\mathcal{F}})_{kj}(X, f_i) \tag{7.12}$$

for all $i, j, k \in \{1, \ldots, n\}$ and arbitrary $X \in T_pM$. The proof is a straightforward computation:

$$\begin{aligned}
(\Omega_{\mathcal{F}})_{ij}(X, f_k) &= g^K(R(X, f_k)f_i, f_j) \\
&= g^K(R(X, f_i)f_k, f_j) \\
&= (\Omega_{\mathcal{F}})_{kj}(X, f_i) .
\end{aligned}$$

Using (7.12) repeatedly, we calculate:

$$(\Omega_{\mathcal{F}})_{ij}(X, f_k) = -(\Omega_{\mathcal{F}})_{ji}(X, f_k)$$
$$= -(\Omega_{\mathcal{F}})_{ki}(X, f_j)$$
$$= (\Omega_{\mathcal{F}})_{ik}(X, f_j)$$
$$= (\Omega_{\mathcal{F}})_{jk}(X, f_i)$$
$$= -(\Omega_{\mathcal{F}})_{kj}(X, f_i)$$
$$= -(\Omega_{\mathcal{F}})_{ij}(X, f_k) \, .$$

This shows that all the $(\Omega_{\mathcal{F}})_{ij}(X, f_k)$ vanish, for an arbitrary $X \in T_p M$. Hence, since all these forms are of type $(1, 1)$ and vanish when one of the arguments is tangent to $\mathcal{F}$, they vanish identically.

This completes the proof of Theorem 7.27. □

As we mentioned before, even when a Künneth structure on a compact manifold M is flat, we do not have control over the global geometry and topology of M, since usually we do not know that the Künneth connection is complete. The Kähler–Künneth case is very different, since in this case the Künneth connection is the Levi–Civita connection of a positive-definite Kähler metric, and so compactness of M does imply completeness via the Hopf–Rinow Theorem. In particular, every compact manifold endowed with a Kähler–Künneth structure is an orientable flat Riemannian manifold, obtained as a compact quotient of Euclidean space by a discrete group of isometries. Therefore, the only candidate in dimension 2 is the torus T^2. However, in dimension 4 the known classification of flat Riemannian Kähler manifolds gives T^4 and seven other candidates, all of which are T^2-bundles over T^2 with first Betti number 2.

Here is a simple example.

Example 7.29 Consider $\varphi = -\operatorname{Id} = \begin{pmatrix} -1 & 0 \\ 0 & -1 \end{pmatrix}$ acting as a linear diffeomorphism on T^2. This is an area-preserving isometry of the standard flat Kähler metric. Moreover, it preserves any pair of linear complementary foliations. Since linear foliations are parallel, φ is an automorphism of a Kähler–Künneth structure. Suspending it as in Subsection 5.3.2 of Chapter 5 using a Kähler–Künneth structure on the base T^2 yields a Kähler–Künneth structure on $M(\varphi) \times S^1$.

The manifold $M(\varphi) \times S^1$ in this example is a flat Riemannian four-manifold with holonomy group $\mathbb{Z}_2$. Similar examples exist in higher dimensions.

One can construct Kähler–Künneth structures by finding Anosov diffeomorphisms of closed flat Kähler manifolds which preserve a Kähler form, and are therefore symplectomorphisms. The Künneth structure induced by the Anosov symplectomorphism together with the Kähler structure then forms a Kähler–Künneth structure. Some concrete examples of this appear in [HK-21].

Notes for Chapter 7

1. Theorem 7.6 goes back to Hess [Hes-80].
2. The discussion in Subsection 7.2.2 goes back to Boyom [Boy-95].
3. Propositions 7.23 and 7.25 are due to Etayo Gordejuela and Santamaría [ES-01]. We gave a proof of Theorem 7.27, without the Künneth terminology, in the context of Kähler geometry in [HK-21], and discovered afterwards that this is also implicit in the discussion on pp. 190–191 of Vaisman's book [Vai-87]. See also Etayo, Santamaría and Trías [EST-06].
4. Example 7.29, showing the existence of a Kähler–Künneth structure on a closed manifold that is not a torus, contradicts a claim made in [Vai-89], see the top of p. 561.

8

Hypersymplectic Geometry

Whenever one is given two non-degenerate two-forms ω and η on the same manifold M, there exists a unique field of invertible endomorphisms A of the tangent bundle TM defined by the equation $i_X\omega = i_{AX}\eta$. The important special case when the two two-forms involved are closed, and therefore symplectic, is very interesting both from the point of view of physics, where it arises in the context of bi-Hamiltonian systems, and from a purely mathematical viewpoint. In physics the field of endomorphisms A is called a recursion operator, and we adopt this terminology here.

In Section 8.1 we consider the simplest examples, where the recursion operator A satisfies $A^2 = \mathrm{Id}$ or $A^2 = -\mathrm{Id}$. We find that these most basic cases correspond precisely to symplectic pairs and to holomorphic symplectic forms respectively. In Section 8.2 we formulate the basics of hypersymplectic geometry in the language of recursion operators. The definition we give is not the original one due to Hitchin [Hit-90] but is equivalent to it. In Section 8.3 we show that every hypersymplectic structure contains a family of Künneth structures parametrised by the circle. The associated metric is independent of the parameter, and is Ricci-flat; cf. Section 8.4.

8.1 Simple Recursion Operators

Consider a pair of symplectic forms (ω, η) on a smooth manifold M.

Definition 8.1 The *recursion operator* defined by the pair (ω, η) is the unique smooth section $A \in \Gamma(\mathrm{End}(TM))$ satisfying the equation

$$i_X\omega = i_{AX}\eta \qquad \forall X, Y \in TM . \tag{8.1}$$

Note that $A(x)$ is invertible for all $x \in M$ because both forms are non-degenerate.

107

We will often encode this definition in the shorthand

$$\eta \xrightarrow{A} \omega \,,$$

deemed to be equivalent to (8.1). This notation is supposed to suggest that applying A to the first argument transforms η into ω. This of course means that A^{-1} transforms ω into η, so that

$$\omega \xrightarrow{A^{-1}} \eta \,.$$

The recursion operator A is the identity if and only if ω and η agree. It is minus the identity if and only if $\omega = -\eta$. From now on we exclude these trivial cases, so we always assume $A \neq \pm \operatorname{Id}$.

When more than two symplectic forms are considered, as will be the case in the definition of hypersymplectic structures, it becomes important not get confused over the composition of recursion operators. We therefore point out that

$$\eta \xrightarrow{A} \omega \xrightarrow{B} \alpha$$

implies

$$\eta \xrightarrow{AB} \alpha \,,$$

with recursion operator AB, not BA.

8.1.1 Symplectic Pairs

Consider first the case $A^2 = \operatorname{Id}_{TM}$, but $A \neq \pm \operatorname{Id}_{TM}$. This means that A is an almost product structure as discussed in Section 4.4 of Chapter 4.

The eigenvalues of A are ± 1, and

$$X = \frac{1}{2}(X + AX) + \frac{1}{2}(X - AX)$$

is the unique decomposition of an arbitrary tangent vector X into a sum of eigenvectors of A. Thus the eigenspaces of A give a splitting $TM = D_+ \oplus D_-$.

Lemma 8.2 *The eigenspaces $D_\pm$ for the eigenvalues ± 1 are precisely the kernels of $\Omega^\mp = \omega \mp \eta$.*

Proof Let X be an arbitrary tangent vector. Then

$$i_X \Omega^\mp = i_X \omega \mp i_X \eta = i_{AX} \eta \mp i_X \eta = i_{AX \mp X} \eta \,.$$

As η is non-degenerate, the condition $i_X \Omega^\mp = 0$ is equivalent to $AX = \pm X$. □

The dimensions of the kernels of $\Omega^{\mp}$ are semicontinuous, in that each can increase only on a closed subset. However, the lemma shows that if the dimension of the kernel of one of the two forms $\Omega^{\mp}$ jumps up, then the dimension of the kernel of the other one has to decrease. Therefore, the dimensions of the kernels are actually constant on a connected manifold M, so the forms $\Omega^{\mp}$ have constant ranks. Moreover, as the $\Omega^{\mp}$ are closed, their kernel distributions are integrable. Thus the forms $\Omega^{\mp}$ are a symplectic pair in the sense of [BK-06]. The two foliations form a bifoliation with symplectic leaves for ω and η. On D_+ the two symplectic forms coincide, and on D_- each one is the negative of the other.

Conversely, suppose that we have a symplectic pair $\Omega^{\pm}$ on M, i.e. a pair of closed two-forms of constant ranks, whose kernel foliations $\mathcal{F}$ and $\mathcal{G}$ are complementary. Then $\omega = \frac{1}{2}(\Omega^+ + \Omega^-)$ and $\eta = \frac{1}{2}(\Omega^+ - \Omega^-)$ are symplectic forms, and the corresponding recursion operator is $A = \mathrm{Id}_{T\mathcal{G}} \ominus \mathrm{Id}_{T\mathcal{F}}$. Thus $A^2 = \mathrm{Id}_{TM}$. We have proved the following.

Proposition 8.3 *Two symplectic forms ω and η on a connected manifold M whose recursion operator A satisfies $A^2 = \mathrm{Id}_{TM}$ and $A \neq \pm\,\mathrm{Id}_{TM}$ give rise to a symplectic pair $\Omega^{\pm}$, and every symplectic pair $\Omega^{\pm}$ arises in this way.*

Note that in this case the two foliations need not be equidimensional, and all pairs of dimensions $(2k, 2n - 2k)$ can be realised. This is rather different from the case of Künneth structures, where the two isotropic foliations have to be equidimensional.

8.1.2 Holomorphic Symplectic Structures

Throughout this subsection we assume that we have two symplectic forms ω and η on a manifold M of dimension $2n$, such that the recursion operator defined by (8.1) satisfies $A^2 = -\mathrm{Id}_{TM}$. This implies $A^{-1} = -A$, so $i_{AX}\omega = -i_X\eta$.

We will prove the following.

Proposition 8.4 *If the recursion operator A satisfies $A^2 = -\mathrm{Id}_{TM}$, then it defines an integrable complex structure with a holomorphic symplectic form whose real and imaginary parts are ω and η. Every holomorphic symplectic form arises in this way.*

Proof In this case A defines an almost complex structure on M. We extend A complex linearly to the complexified tangent bundle $T_{\mathbb{C}}M = TM \otimes_{\mathbb{R}} \mathbb{C}$. The eigenvalues of A are $\pm i$, and

$$ X = \frac{1}{2}(X - iAX) + \frac{1}{2}(X + iAX) $$

is the unique decomposition of a complex tangent vector X into a sum of eigen-vectors of A. As usual, the eigenspaces of A give a splitting $T_{\mathbb{C}}M = T^{1,0} \oplus T^{0,1}$, where $T^{1,0}$ is the $(+i)$-eigenspace, and $T^{0,1}$ is the $(-i)$-eigenspace. The two are complex conjugates of each other.

Lemma 8.5 *The eigenbundles $T^{0,1}$ and $T^{1,0}$ of A are precisely the kernels of $\Omega = \omega + i\eta$ and of its complex conjugate $\overline{\Omega} = \omega - i\eta$.*

Proof It suffices to prove the statement for the $(-i)$-eigenbundle $T^{0,1}$. The other case then follows by complex conjugation.

Let $X = u + iv$ be a complex tangent vector. Then

$$i_X \Omega = i_u \omega - i_v \eta + i(i_u \eta + i_v \omega) \, .$$

The real part of the equation $i_X \Omega = 0$ is equivalent to its imaginary part, and each is equivalent to $Au = v$, which is obviously equivalent to $X \in T^{0,1}$. □

Now we want to show that the almost complex structure A is in fact in-tegrable. By the Newlander–Nirenberg Theorem it suffices to check that one (and hence both) of the eigenbundles of A is closed under commutation. To do this, suppose X and Y are complex vector fields in $T^{1,0}$, so that $AX = iX$, $AY = iY$. Then, extending $L_X = i_X \circ d + d \circ i_X$ complex linearly to complex tangent vectors, and using the fact that ω and η are closed, we find

$$i_{A[X,Y]}\eta = i_{[X,Y]}\omega = L_X i_Y \omega - i_Y L_X \omega = L_X i_{AY}\eta - i_Y L_{AX}\eta$$
$$= i(L_X i_Y \eta - i_Y L_X \eta) = i_{i[X,Y]}\eta \, .$$

The non-degeneracy of η now implies that $A[X, Y] = i[X, Y]$, so $T^{1,0}$ is closed under commutation.

Thus we have seen that two symplectic forms ω and η whose recursion op-erator satisfies $A^2 = -\mathrm{Id}_{TM}$ give rise to an integrable complex structure, for which $T^{0,1}$ is precisely the kernel of $\Omega = \omega + i\eta$. Thus Ω is a closed form of type $(2, 0)$ and rank n, where n is the complex dimension of M.

Conversely, if a manifold is complex and carries a holomorphic symplectic form, then the real and imaginary parts of this form are real symplectic forms whose recursion operator is just the complex structure.

This completes the proof of Proposition 8.4. □

Note that for a manifold with a holomorphic symplectic form the complex dimension n is even, $\Omega^{n/2}$ is nowhere zero and $\Omega^{(n/2)+1}$ is identically zero. If $n = 2$, the last condition becomes $\Omega^2 = 0$, whose real and imaginary parts lead to $\omega \wedge \omega = \eta \wedge \eta$ and $\omega \wedge \eta = 0$. Thus ω and η precisely form a conformal symplectic couple in the sense of Geiges [Gei-96].

8.2 Hypersymplectic Structures

We now consider a geometric structure defined by a triple of symplectic forms whose pairwise recursion operators all have squares equal to $\pm$ Id. Depending on the choices of signs, there are, up to permutation, three other structures of this type, but we will not consider those here. We refer the interested reader to Bande and Kotschick [BK-08] for a discussion of those structures.

Definition 8.6   A *hypersymplectic structure* on a smooth manifold M consists of a triple of symplectic forms ω, α and β whose recursion operators specified by the commutative diagram

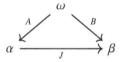

satisfy $A^2 = B^2 = \mathrm{Id}_{TM} = -J^2$.

The diagram must commute since AJ and B are both recursion operators transforming ω into β, and, by the non-degeneracy of the symplectic forms, the recursion operators are uniquely determined, so that $AJ = B$. Note that in this situation it cannot happen that A or B is $\pm \mathrm{Id}_{TM}$, since this would lead to J being, up to sign, the same as B or A, which would be inconsistent with the assumption $A^2 = B^2 = -J^2$. Thus both A and B are non-trivial almost product structures.

By our discussion in the previous section, a hypersymplectic structure makes M into a holomorphic symplectic manifold since J is integrable, and $\beta + i\alpha$ is holomorphic symplectic with respect to J. Moreover, we have two different symplectic pairs given by $\omega \pm \alpha$ and $\omega \pm \beta$. In order to be able to analyse the interplay of these structures forced by the hypersymplectic definition, we first note some purely algebraic consequences of that definition.

Lemma 8.7   *For any hypersymplectic structure the three recursion operators pairwise anti-commute.*

Proof   We have $AJ = B$ by definition. Inverting, we find $J^{-1}A^{-1} = B^{-1}$. Now $J^{-1} = -J$, whereas A and B are their own inverses, so this equation becomes $-JA = B$, showing that A and J anti-commute.

We also have

$$BJ = (AJ)J = AJ^2 = -A$$

and, using the anti-commuting property of A and J,

$$JB = J(AJ) = -J(JA) = -J^2A = A \ ,$$

showing that B and J anti-commute.

Finally,

$$AB = A(AJ) = A^2 J = J \,,$$

whereas, using again that A and J anti-commute,

$$BA = (AJ)A = -A^2 J = -J \,.$$

This shows that A and B also anti-commute. $\square$

Now these algebraic identities have the following important consequence.

Proposition 8.8 *For any hypersymplectic structure ω, α, β we have*

$$\omega(X, JY) = \alpha(X, BY) = -\beta(X, AY) \,, \tag{8.2}$$

and these expressions define a bilinear form $g(X, Y)$ on TM. It is non-degenerate and symmetric, is invariant under J and satisfies $g(AX, AY) = g(BX, BY) = -g(X, Y)$.

Proof To prove the first equality in (8.2) we calculate using the relations between the recursion operators and the skew-symmetry of the symplectic forms:

$$\omega(X, JY) = \omega(X, ABY) = -\omega(ABY, X) = -\alpha(BY, X) = \alpha(X, BY) \,.$$

Similarly for the second equality in (8.2):

$$\alpha(X, BY) = \alpha(X, AJY) = \alpha(JAY, X) = \beta(AY, X) = -\beta(X, AY) \,.$$

We have now proved (8.2).

We define $g(X, Y) = \omega(X, JY)$. Since J is an isomorphism and ω is non-degenerate, it follows that the bilinear form g is also non-degenerate. Next we check that g is symmetric using (8.2):

$$g(Y, X) = \alpha(Y, BX) = \omega(AY, BX) = -\beta(X, AY) = g(X, Y) \,.$$

Using the symmetry of g and the skew-symmetry of ω we find

$$g(JX, JY) = \omega(JX, J^2 Y) = \omega(Y, JX) = g(Y, X) = g(X, Y) \,,$$

showing that g is indeed J-invariant. In the same way we obtain

$$g(AX, AY) = -\beta(AX, A^2 Y) = \beta(Y, AX) = -g(Y, X) = -g(X, Y)$$

and

$$g(BX, BY) = \alpha(BX, B^2 Y) = -\alpha(Y, BX) = -g(Y, X) = -g(X, Y) \,.$$

This completes the proof. $\square$

We have shown that a hypersymplectic structure gives rise to an intrinsically defined pseudo-Riemannian metric g. By non-degeneracy, in each tangent space there must exist a vector X with $g(X, X) \neq 0$. We then have $g(JX, JX) = g(X, X)$, and $g(AX, AX) = g(BX, BX) = -g(X, X)$. Furthermore, X, JX, AX and BX are pairwise g-orthogonal to each other. Thus they span a four-dimensional subspace on which g is non-degenerate and has signature $(2, 2)$. Looking at the orthogonal complement of this subspace and proceeding inductively, we see that the metric g has neutral signature.

Definition 8.6 and the above elementary considerations can be used to give a direct proof of the following.

Proposition 8.9 *For any hypersymplectic structure the recursion operators commute with the Levi–Civita connection of the associated neutral metric g. Furthermore, the Levi–Civita connection is compatible with the three symplectic forms.*

We do not give the proof here, but postpone it to the next section, where it will follow automatically from our discussion of the Künneth connection. This proposition implies that the neutral metric of a hypersymplectic structure has reduced holonomy, with the reduction specified by the parallel endomorphisms A, B and J. One can turn this holonomy reduction into an equivalent definition of hypersymplectic structures, and this was one of the viewpoints explained by Hitchin in [Hit-90]. He paraphrased the definition in terms of holonomy by defining a hypersymplectic structure on a manifold M to be a pseudo-Riemannian metric g of neutral signature, together with three endomorphisms I, S and T of the tangent bundle satisfying

$$I^2 = -\operatorname{Id}_{TM}, \quad S^2 = T^2 = \operatorname{Id}_{TM}, \quad IS = -SI = T,$$

$$g(IX, IY) = g(X, Y), \quad g(SX, SY) = -g(X, Y), \quad g(TX, TY) = -g(X, Y),$$

and such that the following three two-forms are closed:

$$\omega_I(X, Y) = g(IX, Y), \quad \omega_S(X, Y) = g(SX, Y), \quad \omega_T(X, Y) = g(TX, Y).$$

In this situation, the three closed two-forms are symplectic, since g is non-degenerate and the endomorphisms are invertible. Given a hypersymplectic structure in this sense, the recursion operators intertwining the three symplectic forms are given by the diagram

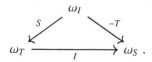

Since $S^2 = (-T)^2 = \mathrm{Id}_{TM} = -I^2$ by definition, it follows that a hypersymplectic structure in Hitchin's sense does give us a hypersymplectic structure in the sense of Definition 8.6. Conversely, given a structure as in Definition 8.6, we have seen that there is an associated pseudo-Riemannian metric g of neutral signature, such that all the requirements of Hitchin's definition are satisfied. Thus the definition given here is indeed equivalent to Hitchin's definition in [Hit-90].

As noted before, by the discussion in Subsection 8.1.2 above, the almost complex structure J that is part of a hypersymplectic structure is automatically integrable, endowing M with a complex manifold structure for which $\beta + i\alpha$ is holomorphic symplectic. Therefore the complex dimension n of M is even, and the real dimension $2n$ is a multiple of 4. This also follows from our discussion of the signature of g.

Examples of hypersymplectic structures on nilmanifolds have appeared in many papers; see for instance [FPPS-04, AD-06]. We will give two examples in Section 9.5 of Chapter 9. We will give the classification of compact four-dimensional hypersymplectic manifolds in Proposition 10.11 of Chapter 10.

8.3 An S^1-Family of Künneth Structures

By its very definition, a hypersymplectic structure includes two symplectic pairs $\omega \pm \alpha$ and $\omega \pm \beta$ in the sense of Subsection 8.1.1 above. The kernel foliation of $\omega \pm \alpha$ has symplectic leaves in the sense that ω and α restrict to symplectic forms on the leaves, they agree on the kernel of $\omega - \alpha$ and they are negatives of each other on the kernel of $\omega + \alpha$. However, the leaves of these kernel foliations are Lagrangian for β and therefore form a Künneth structure. To see this, recall from Subsection 8.1.1 above that the kernel foliations of $\omega \pm \alpha$ have as tangent subbundles the eigenbundles of the corresponding recursion operator intertwining ω and α, which is A. So suppose for example that X and Y are both in the $(+1)$-eigenbundle for A. Then using Proposition 8.8 we find

$$\beta(X, Y) = \beta(X, AY) = -g(X, Y)$$
$$= g(AX, AY) = -\beta(AX, A^2 Y) = -\beta(X, Y),$$

showing that the $(+1)$-eigenbundle for A is indeed isotropic for β. The same argument also works for the (-1)-eigenbundle of A.

We can embed this Künneth structure in a whole family of Künneth structures parametrised by the circle, by treating A and B symmetrically, and taking suitable linear combinations of them.

Given a hypersymplectic structure ω, α and β on M with recursion operators A, B and J, we consider the family of two-forms

$$\alpha_\theta = \cos(\theta)\alpha + \sin(\theta)\beta$$

on M depending on a real parameter θ and the family of endomorphisms of the tangent bundle TM given by

$$I_\theta = \cos(\theta)A + \sin(\theta)B .$$

We now prove that these data define an S^1-family of Künneth structures.

Theorem 8.10 *The forms α_θ are symplectic and the endomorphisms I_θ are integrable almost product structures for all θ. Moreover, for every θ the symplectic form $\alpha_{\theta+\pi/2}$ together with the eigenfoliations $\mathcal{F}_\theta$ and $\mathcal{G}_\theta$ of I_θ form a Künneth structure.*

Proof Consider first the endomorphisms I_θ and compute the square:

$$I_\theta^2 = \cos^2(\theta)A^2 + \cos(\theta)\sin(\theta)(AB + BA) + \sin^2(\theta)B^2 .$$

Now $A^2 = B^2 = \mathrm{Id}_{TM}$ and the fact that A and B anti-commute tell us that $I_\theta^2 = \mathrm{Id}_{TM}$ for all θ.

Next, each α_θ is a closed two-form on M. Moreover, we have $\omega \xrightarrow{I_\theta} \alpha_\theta$, and so the non-degeneracy of ω and the invertibility of I_θ imply that all α_θ are non-degenerate and therefore symplectic.

The eigenbundles of the almost product structure I_θ are the kernels of the closed forms $\omega \pm \alpha_\theta$, and are therefore integrable to foliations $\mathcal{F}_\theta$ and $\mathcal{G}_\theta$. It remains to show that these foliations are Lagrangian for $\alpha_{\theta+\pi/2}$.

Observe that

$$I_{\theta+\pi/2} = \cos(\theta + \pi/2)A + \sin(\theta + \pi/2)B = -\sin(\theta)A + \cos(\theta)B .$$

Using this we can compute the composition

$$
\begin{aligned}
I_\theta I_{\theta+\pi/2} &= (\cos(\theta)A + \sin(\theta)B)(-\sin(\theta)A + \cos(\theta)B) \\
&= \sin(\theta)\cos(\theta)(B^2 - A^2) + \cos^2(\theta)AB - \sin^2(\theta)BA \\
&= AB ,
\end{aligned}
$$

where in the last step we used $B^2 = A^2$ and the fact that A and B anti-commute. In the same way one computes

$$I_{\theta+\pi/2}I_\theta = BA ,$$

showing that I_θ and $I_{\theta+\pi/2}$ anti-commute because A and B do so. This means that $I_{\theta+\pi/2}$ interchanges the eigenbundles of I_θ, which are therefore of equal rank. In other words, $\mathcal{F}_\theta$ and $\mathcal{G}_\theta$ are equidimensional foliations.

Suppose now that X and Y are in the $(+1)$-eigenbundle of I_θ. Then writing

$$\begin{aligned}
\alpha_{\theta+\pi/2}(X, Y) &= \omega(I_{\theta+\pi/2}X, Y) \\
&= \frac{1}{2}(\omega + \alpha_\theta)(I_{\theta+\pi/2}X, Y) + \frac{1}{2}(\omega - \alpha_\theta)(I_{\theta+\pi/2}X, Y) ,
\end{aligned}$$

the first summand on the right-hand side vanishes because $I_{\theta+\pi/2}X$ is in the (-1)-eigenbundle of I_θ, which is the kernel of $\omega + \alpha_\theta$. The second summand on the right-hand side vanishes because Y is in the $(+1)$-eigenbundle of I_θ, which is the kernel of $\omega - \alpha_\theta$.

We conclude that the $(+1)$-eigenbundle of I_θ is isotropic for $\alpha_{\theta+\pi/2}$. The same argument works for the (-1)-eigenbundle, so the two foliations $\mathcal{F}_\theta$ and $\mathcal{G}_\theta$ tangent to these eigenbundles are indeed Lagrangian for $\alpha_{\theta+\pi/2}$. This completes the proof of the theorem. $\qquad\square$

We now see that any hypersymplectic structure comes embedded in a whole S^1-family of such structures

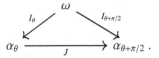

The S^1-family of Künneth structures naturally arises from this S^1-family of hypersymplectic structures.

Remark 8.11 We can describe the (± 1)-eigenbundles of I_θ explicitly in terms of the (± 1)-eigenbundles of A. Since $B = AJ = -JA$, we can write

$$I_\theta = Ae^{\theta J} = e^{-\theta J}A ,$$

where

$$e^{\theta J} = \sum_{n=0}^{\infty} \frac{(\theta J)^n}{n!} = \cos(\theta)\mathrm{Id}_{TM} + \sin(\theta)J .$$

It follows that

$$I_\theta X = \pm X \Leftrightarrow Ae^{\theta J}X = \pm X \Leftrightarrow e^{-\frac{1}{2}\theta J}Ae^{\frac{1}{2}\theta J}X = \pm X$$

$$\Leftrightarrow Ae^{\frac{1}{2}\theta J}X = \pm e^{\frac{1}{2}\theta J}X .$$

Hence if $Y \in TM$ satisfies

$$AY = \pm Y ,$$

then

$$X = e^{-\frac{1}{2}\theta J} Y = \cos\left(\frac{\theta}{2}\right) Y - \sin\left(\frac{\theta}{2}\right) JY$$

satisfies

$$I_\theta X = \pm X .$$

Recall that for every almost Künneth structure there is an associated neutral metric g, well defined up to sign. If the almost Künneth structure is given by (ω, F, G), then g is defined by

$$g(X, Y) = \omega(IX, Y)$$

with $I = \mathrm{Id}_F \ominus \mathrm{Id}_G$. For the Künneth structures arising from a hypersymplectic structure, this means that

$$g(X, Y) = \alpha_{\theta+\pi/2}(I_\theta X, Y) = \omega(I_{\theta+\pi/2} I_\theta X, Y) = \omega(BAX, Y) = -\omega(JX, Y) .$$

Comparing with (8.2) in Proposition 8.8, we see that this is precisely the metric defined by the hypersymplectic structure. We have proved the following.

Proposition 8.12 *For every θ, the neutral metric defined by the Künneth structure $(\alpha_{\theta+\pi/2}, \mathcal{F}_\theta, \mathcal{G}_\theta)$ equals the hypersymplectic metric g defined in Proposition 8.8.*

This proposition together with Theorem 6.10 imply the following:

Corollary 8.13 *For every θ, the Künneth connection of the Künneth structure $(\alpha_{\theta+\pi/2}, \mathcal{F}_\theta, \mathcal{G}_\theta)$ is the Levi–Civita connection of the hypersymplectic metric g defined in Proposition 8.8.*

We can now supply the proof we postponed in the previous section.

Proof of Proposition 8.9 Proposition 8.9 stated that for any hypersymplectic structure the three recursion operators commute with the Levi–Civita connection of the associated neutral metric g. Since g is parallel for its own Levi–Civita connection, the relations between g and the symplectic forms given in Proposition 8.8 then imply that the three symplectic forms are parallel.

Now the Levi–Civita connection of g is the Künneth connection for the Künneth structures $(\alpha_{\theta+\pi/2}, \mathcal{F}_\theta, \mathcal{G}_\theta)$, and so it commutes with I_θ for all θ. In particular, it commutes with A and B. Therefore it also commutes with $J = AB$. □

So, as θ varies in this family of Künneth structures, the neutral metric g is independent of θ, and therefore its Levi–Civita connection, which is the Künneth connection according to Theorem 6.10, is also independent of θ. The Künneth connection is compatible with the symplectic form of a Künneth structure, and here this means that the Künneth connection, which is independent of θ, is compatible with α_θ for all θ. The eigenfoliations of I_θ are Lagrangian for $\alpha_{\theta+\pi/2}$, but they are symplectic for α_θ. Thus the leaves of these foliations carry symplectic forms that are parallel for the restriction of the Künneth connection to the leaves, which we know to be the flat Bott connection. So the leaves are not just affinely flat – which is true for all Lagrangian foliations – but they are affinely flat and symplectic, with symplectic forms that are compatible with the flat affine structure. The top power of the symplectic form on a leaf is then a parallel volume form, and so the leaves come equipped with volume forms parallel with respect to their flat affine structure. This means that, by the recent result of Klingler [Kli-17], we obtain the following.

Theorem 8.14   *Any closed leaf of one of the foliations underlying a hypersymplectic structure has vanishing Euler characteristic.*

8.4 The Ricci Curvature of Hypersymplectic Structures

In this section we prove that hypersymplectic manifolds are Ricci-flat and are thus examples of neutral Calabi–Yau manifolds. The proof proceeds by showing that the Ricci curvature is essentially the curvature of the induced connection on the canonical bundle. This proves the vanishing of the Ricci curvature because the canonical bundle is trivialised by the top power of a holomorphic symplectic form.

As before, suppose that

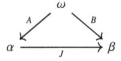

is a hypersymplectic structure on a manifold M of real dimension $4m$. According to our discussion in Subsection 8.1.2 above, J makes M into a complex manifold of complex dimension $2m$ endowed with the holomorphic symplectic form $\beta + i\alpha$. We have seen in Proposition 8.9 that α and β are parallel for the Levi–Civita connection of the hypersymplectic metric. Therefore $(\beta + i\alpha)^m$ is a parallel section – with respect to the induced connection – of the canonical bundle $K_M = \Lambda^{2m,0}(M)$. This shows that the Levi–Civita connection induces a flat connection on the canonical bundle. Therefore, to show Ricci-flatness, it suffices to show that the Ricci curvature is essentially the curvature of the canonical bundle. This is a standard argument in Kähler geometry that we now carry out in the case of a pseudo-Kähler structure of neutral signature.

We can calculate the trace involved in the Ricci curvature most easily in a local frame. Let $e_1, \ldots, e_{2m}$ be local orthogonal vector fields, so that the vectors

$$e_1, Je_1, \ldots, e_{2m}, Je_{2m}$$

form a local orthogonal basis for TM. We can choose the vectors such that

$$g(e_k, e_k) = g(Je_k, Je_k) = +1 \quad \forall k = 1, \ldots, m,$$
$$g(e_k, e_k) = g(Je_k, Je_k) = -1 \quad \forall k = m+1, \ldots, 2m.$$

We define dual local one-forms $\tau_1, \ldots, \tau_{2m}$ on M by

$$\tau_j(e_k) = \delta_{jk},$$
$$\tau_j(Je_k) = i\delta_{jk}.$$

Then the one-forms τ_j are complex linear and form a local basis for $\Lambda^{1,0}(M)$. This implies that

$$\mu = \tau_1 \wedge \ldots \wedge \tau_{2m}$$

defines a local trivialisation for the canonical bundle.

Lemma 8.15 *We can write the forms τ_j as*

$$\tau_j(X) = g(X, e_j) + ig(X, Je_j) \quad \forall j = 1, \ldots, m$$

and

$$\tau_j(X) = -g(X, e_j) - ig(X, Je_j) \quad \forall j = m, \ldots, 2m.$$

Another direct calculation using the definition of the curvature on forms shows the following.

Lemma 8.16 *The one-forms $R(X, Y)\tau_j$ are of type $(1,0)$ and*

$$(R(X, Y)\tau_j)(e_k) = -\tau_j(R(X, Y)e_k)$$

for all $k = 1, \ldots, 2m$.

Proposition 8.17 *The curvature $R(X, Y)$ acting on the canonical bundle is given by*

$$R(X, Y)\mu = -ir(X, Y)\mu,$$

where

$$r(X, Y) = \sum_{k=1}^{m} g(R(X, Y)e_k, Je_k) - \sum_{k=m+1}^{2m} g(R(X, Y)e_k, Je_k).$$

Proof We have

$$R(X, Y)\mu = \sum_{k=1}^{2m} \tau_1 \wedge \tau_{k-1} \wedge (R(X, Y)\tau_k) \wedge \tau_{k+1} \ldots \wedge \tau_{2m}.$$

Evaluating both sides on $e_1, \ldots, e_{2m}$, we get

$$(R(X, Y)\mu)(e_1, \ldots, e_{2m}) = \sum_{k=1}^{2m} (R(X, Y)\tau_k)(e_k) .$$

We have

$$\sum_{k=1}^{2m} (R(X, Y)\tau_k)(e_k) = -\sum_{k=1}^{2m} \tau_k(R(X, Y)e_k)$$

$$= -\sum_{k=1}^{m} (g(R(X, Y)e_k, e_k) + ig(R(X, Y)e_k, Je_k))$$

$$+ \sum_{k=m+1}^{2m} (g(R(X, Y)e_k, e_k) + ig(R(X, Y)e_k, Je_k)) .$$

However, for the curvature of a pseudo-Riemannian metric,

$$g(R(X, Y)V, V) = 0 ,$$

for arbitrary vectors X, Y, V, hence the statement follows.   □

Proposition 8.18 *The Ricci curvature of the hypersymplectic metric g is given by*

$$\mathrm{Ric}(X, Y) = r(JX, Y) .$$

Proof By definition, the Ricci curvature of the pseudo-Riemannian metric g can be calculated as

$$\mathrm{Ric}(X, Y) = \sum_{k=1}^{m} (g(R(e_k, X)Y, e_k) + g(R(Je_k, X)Y, Je_k))$$

$$- \sum_{k=m+1}^{2m} (g(R(e_k, X)Y, e_k) + g(R(Je_k, X)Y, Je_k)) .$$

Using the symmetries of the curvature and that J is compatible with the Levi–

Civita connection we get

$$
\begin{aligned}
\text{Ric}(X, Y) &= \sum_{k=1}^{m} (g(R(e_k, X)JY, Je_k) - g(R(Je_k, X)JY, e_k)) \\
&\quad - \sum_{k=m+1}^{2m} (g(R(e_k, X)JY, Je_k) - g(R(Je_k, X)JY, e_k)) \\
&= \sum_{k=1}^{m} (g(R(X, e_k)Je_k, JY) + g(R(Je_k, X)e_k, JY)) \\
&\quad - \sum_{k=m+1}^{2m} (g(R(X, e_k)Je_k, JY) + g(R(Je_k, X)e_k, JY)) \\
&= -\sum_{k=1}^{m} g(R(e_k, Je_k)X, JY) + \sum_{k=m+1}^{2m} g(R(e_k, Jc_k)X, JY) \\
&= \sum_{k=1}^{m} g(R(e_k, Je_k)JX, Y) - \sum_{k=m+1}^{2m} g(R(e_k, Je_k)JX, Y) \\
&= \sum_{k=1}^{m} g(R(JX, Y)e_k, Je_k) - \sum_{k=m+1}^{2m} g(R(JX, Y)e_k, Je_k) \\
&= r(JX, Y).
\end{aligned}
$$

This completes the proof. $\qquad\qquad\qquad\qquad\qquad\qquad\qquad\qquad\quad$ □

We can now finally prove the Ricci-flatness of hypersymplectic structures.

Theorem 8.19 *The neutral metric associated with a hypersymplectic structure is Ricci-flat.*

Proof By Proposition 8.18, Ricci-flatness is equivalent to the vanishing of the tensor r. By Proposition 8.17, the tensor r is the curvature tensor of the induced connection on the canonical bundle. Hence it vanishes because the canonical bundle has a non-vanishing parallel section. $\qquad\qquad\qquad\qquad$ □

According to Corollary 8.13, this theorem tells us that all the Künneth structures $(\alpha_{\theta+\pi/2}, \mathcal{F}_\theta, \mathcal{G}_\theta)$ derived from a hypersymplectic structure are Ricci-flat, and thus in particular are Künneth–Einstein structures.

Notes for Chapter 8

1. Symplectic pairs were introduced and studied by Bande and Kotschick in [BK-06].
2. In Sections 8.1 and 8.2 we follow Bande and Kotschick [BK-08]. Proposition 8.8 corrects a sign in [BK-08, Proposition 16].
3. Ricci-flatness of hypersymplectic structures is often asserted; see, for example, [Hit-90, DS-05], but usually not proved in the literature. See also [FPPS-04].
4. We refer the reader to the survey by Dancer and Swann [DS-05] for other results and a different point of view on hypersymplectic manifolds.

9
Nilmanifolds

This chapter contains a brief introduction to nilmanifolds, and a discussion of Künneth and related structures on nilmanifolds. Nilmanifolds are homogeneous spaces for nilpotent Lie groups, and for them the discussions of geometric structures can often be reduced to the consideration of left-invariant structures. Left-invariant structures in turn arise from the corresponding linear structures on the Lie algebra, and these linear structures are usually much more tractable than arbitrary geometric structures on smooth manifolds. The nilmanifolds of abelian Lie groups are just tori, so in some sense nilmanifolds are the simplest generalisations of tori.

We do not give a systematic treatment of nilmanifolds here, but focus on providing a few explicit examples of Künneth structures, hypersymplectic structures and Anosov symplectomorphisms in this setting. For more information on topics from the theory of nilmanifolds that we treat rather breezily, we refer the reader to the books by Gorbatsevich, Onishchik and Vinberg [GOV-97] and by Knapp [Kna-96].

9.1 Nilpotent and Solvable Lie Algebras

We begin with some standard algebraic background on nilpotent and solvable Lie algebras. Let L be a (real or complex) Lie algebra with Lie bracket $[\cdot, \cdot]$.

Definition 9.1 For subsets $A, B \subset L$ we define the vector subspace $[A, B] \subset L$ as the set of finite sums of elements of the form $[a, b]$, where $a \in A$ and $b \in B$. A linear subspace $I \subset L$ such that $[L, I] \subset I$ is called an *ideal* in L. A linear subspace $A \subset L$ such that $[A, A] \subset A$ is called a *Lie subalgebra*.

The consideration of iterated commutators leads to two different sequences of subspaces, which we now define.

Definition 9.2   The *lower central series* of L is defined inductively for positive integers i:

$$L_{(1)} = L ,$$
$$L_{(i)} = [L, L_{(i-1)}] \quad \forall i \geq 2 .$$

By induction it follows easily that

$$L = L_{(1)} \supset L_{(2)} \supset L_{(3)} \supset \dots$$

In particular, $L_{(i)}$ is an ideal in L for all $i \geq 1$.

We will be particularly interested in Lie algebras L that have the following property.

Definition 9.3   The Lie algebra L is called *nilpotent* if $L_{(r+1)} = 0$ for some integer $r \geq 0$. If r is the smallest such integer, then L is called *r-step nilpotent*. Hence $r = 0$ if and only if $L = 0$, and if $r > 0$, then $L_{(r)} \neq 0$ and $L_{(r+1)} = 0$.

This immediately implies the following.

Proposition 9.4   *Abelian Lie algebras are nilpotent. More precisely, the non-zero abelian Lie algebras are precisely the one-step nilpotent Lie algebras (and the zero Lie algebra is the zero-step nilpotent Lie algebra).*

We can modify the definition above and consider the following sequence of subspaces of a Lie algebra L.

Definition 9.5   The *derived series* of L is defined inductively for non-negative integers i:

$$L^{(0)} = L ,$$
$$L^{(i)} = \left[L^{(i-1)}, L^{(i-1)} \right] \quad \forall i \geq 1 .$$

It follows easily that

$$L = L^{(0)} \supset L^{(1)} \supset L^{(2)} \supset \dots$$

The linear subspace $L^{(i)}$ is a Lie subalgebra of L for all integers $i \geq 0$.

Definition 9.6   The Lie algebra L is called *solvable* if $L^{(s)} = 0$ for some integer s.

Note that by convention the lower central series starts at index 1 and the derived series starts at index 0. The lower central series is a sequence of ideals and the derived series is a sequence of Lie subalgebras of L. We want to discuss some simple algebraic properties of the lower central and of the derived series.

Lemma 9.7 *Let L be a Lie algebra.*

(i) *The lower central series satisfies* $\left[L_{(i)}, L_{(j)}\right] \subset L_{(i+j)}$ *for all* $i, j \geq 1$.
(ii) *The lower central and the derived series are related by* $L^{(k)} \subset L_{(2^k)}$ *for all* $k \geq 0$.

Proof Both statements are proved by induction.

For the first statement we fix i and do induction on j. The claim is true for $j = 1$. If we assume that the claim is true for some specific $j \geq 1$ and all i, then by the Jacobi identity and the definition of the lower central series we find

$$\left[L_{(i)}, L_{(j+1)}\right] = \left[L_{(i)}, \left[L_{(j)}, L\right]\right]$$
$$\subset \left[L_{(j)}, [L, L_{(i)}]\right] + \left[L, \left[L_{(i)}, L_{(j)}\right]\right]$$
$$\subset L_{(i+j+1)} .$$

For the proof of the second statement we use induction on k. The claim is true for $k = 0$. If we assume that the claim is true for some $k \geq 0$, then

$$L^{(k+1)} = \left[L^{(k)}, L^{(k)}\right] \subset \left[L_{(2^k)}, L_{(2^k)}\right] \subset L_{(2^{k+1})}$$

by the first part of the lemma. □

The following proposition is a simple consequence of Lemma 9.7 (ii).

Proposition 9.8 *Nilpotent Lie algebras are solvable.*

We thus have the inclusions of Lie algebras

$$\{\text{abelian}\} \subset \{\text{nilpotent}\} \subset \{\text{solvable}\} . \tag{9.1}$$

As we shall see later, there is a corresponding sequence of inclusions of closed manifolds

$$\{\text{tori}\} \subset \{\text{nilmanifolds}\} \subset \{\text{solvmanifolds}\} .$$

We want to consider two simple classes of examples of nilpotent and solvable Lie algebras. This will show that the inclusions in (9.1) are strict.

Example 9.9 Let $\mathbb{K} = \mathbb{R}$ or $\mathbb{C}$ and consider for $k \geq 1$ the Lie algebra

$$\mathfrak{gl}(k, \mathbb{K}) = \mathbb{K}^{k \times k}$$

with the commutator defined as the commutator of matrix multiplication. The vector subspace

$$\mathfrak{n}_k = \left\{A \in \mathbb{K}^{k \times k} \mid A \text{ is strictly upper-triangular}\right\}$$

is a nilpotent Lie subalgebra of $\mathfrak{gl}(k, \mathbb{K})$, but is not abelian for $k \geq 3$. For example, for $k = 3$ the matrices in $\mathfrak{n}_3$ are of the form

$$\begin{pmatrix} 0 & * & * \\ 0 & 0 & * \\ 0 & 0 & 0 \end{pmatrix}.$$

Then $[\mathfrak{n}_3, \mathfrak{n}_3]$ consists of matrices of the form

$$\begin{pmatrix} 0 & 0 & * \\ 0 & 0 & 0 \\ 0 & 0 & 0 \end{pmatrix}$$

and $[\mathfrak{n}_3, [\mathfrak{n}_3, \mathfrak{n}_3]] = 0$. Hence $\mathfrak{n}_3$ is two-step nilpotent.

Example 9.10   Similarly, let

$$\mathfrak{b}_k = \left\{ A \in \mathbb{K}^{k \times k} \mid A \text{ is upper-triangular} \right\}.$$

Then $\mathfrak{b}_k$ is a solvable Lie subalgebra of $\mathfrak{gl}(k, \mathbb{K})$ but is not nilpotent if $k \geq 2$. For example $\mathfrak{b}_2$ consists of matrices of the form

$$\begin{pmatrix} * & * \\ 0 & * \end{pmatrix},$$

and $[\mathfrak{b}_2, \mathfrak{b}_2]$ consists of matrices of the form

$$\begin{pmatrix} 0 & * \\ 0 & 0 \end{pmatrix},$$

and finally $[[\mathfrak{b}_2, \mathfrak{b}_2], [\mathfrak{b}_2, \mathfrak{b}_2]] = 0$, hence $\mathfrak{b}_2$ is solvable. However, $[\mathfrak{b}_2, [\mathfrak{b}_2, \mathfrak{b}_2]] = [\mathfrak{b}_2, \mathfrak{b}_2]$, so the lower central series stabilises. Therefore $\mathfrak{b}_2$ is not nilpotent.

It is not difficult to show the following (see [Kna-96, Chapter I, Proposition 1.10]).

Proposition 9.11   *Any Lie subalgebra of a nilpotent (respectively solvable) Lie algebra is nilpotent (respectively solvable).*

9.2 Nil- and Infra-nilmanifolds

For a Lie group G we usually denote its Lie algebra by $\mathfrak{g}$.

Definition 9.12   A connected Lie group G is called *nilpotent*, respectively *solvable*, if its Lie algebra $\mathfrak{g}$ is nilpotent, respectively solvable. A nilpotent Lie group G is called *r-step nilpotent* if its Lie algebra $\mathfrak{g}$ is r-step nilpotent.

We will first focus on nilpotent Lie groups and postpone the discussion of the solvable case until Section 9.7 below.

Example 9.13   The simply connected nilpotent Lie group N_k associated to the real nilpotent Lie algebra $\mathfrak{n}_k$ from Example 9.9 consists of the upper triangular matrices $M \in \mathbb{R}^{k \times k}$ of the form

$$
M = \begin{pmatrix} 1 & * & * & \ldots & * \\ 0 & 1 & * & \ldots & * \\ 0 & 0 & 1 & \ldots & * \\ \vdots & & & & \vdots \\ 0 & \ldots & \ldots & \ldots & 1 \end{pmatrix}.
$$

It is diffeomorphic to the Euclidean space $\mathbb{R}^{k(k-1)/2}$.

A fundamental theorem concerning nilpotent Lie groups is the following.

Theorem 9.14   *Let G be a nilpotent Lie group. Then the exponential map*

$$
\exp \colon \mathfrak{g} \longrightarrow G
$$

is a covering map. In particular, if G is simply connected and n-dimensional, then the exponential map is a diffeomorphism of $\mathfrak{g} \cong \mathbb{R}^n$ onto G. Its smooth inverse is denoted by

$$
\log \colon G \longrightarrow \mathfrak{g} .
$$

A proof of this theorem can be found in [Kna-96, Chapter I, Theorem 1.104]. Given a nilpotent Lie group we can always consider its universal covering Lie group, which according to this theorem is diffeomorphic to a Euclidean space via the exponential map.

To motivate the next definition recall how tori can be defined as smooth manifolds. One starts with Euclidean space $\mathbb{R}^n$, which has the structure of a simply connected abelian Lie group under vector addition. In $\mathbb{R}^n$ one considers a subgroup $\mathbb{Z}^n$, called a lattice, and forms the quotient $T^n = \mathbb{R}^n / \mathbb{Z}^n$ under the natural right (or left) action of $\mathbb{Z}^n$ on $\mathbb{R}^n$. The torus T^n is a closed manifold.

This construction generalises to nilpotent Lie groups as follows.

Definition 9.15   Let G be a simply connected nilpotent Lie group. A *lattice* $\Gamma \subset G$ is a discrete subgroup that is *co-compact* (also called *uniform*), i.e. the quotient $M = \Gamma \backslash G$ is compact. The topological space M has an induced structure of a smooth manifold and is called a *nilmanifold*.

Here the quotient is taken with respect to the canonical left action of Γ on G. In the special case where G is abelian, the definition of nilmanifolds reduces to the definition of tori. Since simply connected nilpotent Lie groups are diffeomorphic to $\mathbb{R}^n$, nilmanifolds M are quotients of $\mathbb{R}^n$ by a left action of a discrete, nilpotent group Γ. In particular, the fundamental group of M is isomorphic to Γ.

A nilmanifold has a canonical structure as a homogeneous G-space, because there exists a transitive, smooth right G-action on $\Gamma \backslash G$. According to a theorem of Malcev [Malc-62] the converse also holds; every compact manifold that admits a transitive, smooth action of a nilpotent Lie group is a nilmanifold.

It turns out that not every simply connected nilpotent Lie group has a lattice. However, the following theorem, also due to Malcev [Malc-62], gives a necessary and sufficient criterion for the existence of lattices.

Theorem 9.16 (Malcev [Malc-62]) *A simply connected nilpotent Lie group G has a lattice if and only if its Lie algebra $\mathfrak{g}$ admits a basis with rational structure constants.*

The structure constants with respect to a basis $e_1, \ldots, e_n$ of $\mathfrak{g}$ are given by the coefficients c_{ij}^k in the expansion

$$\left[e_i, e_j \right] = \sum_{k=1}^{n} c_{ij}^k e_k .$$

The idea of the proof of Malcev's Theorem is that a basis of $\mathfrak{g}$ with rational structure constants determines via the exponential map a lattice in G. Conversely, a lattice in G determines via the logarithm map a basis of $\mathfrak{g}$ with rational structure constants.

Example 9.17   We consider the nilpotent Lie algebra $\mathfrak{nil}_3 = \mathfrak{n}_3$ as in Example 9.9 with simply connected nilpotent Lie group $Nil^3 = N_3$. The elements of Nil^3 are matrices of the form

$$\begin{pmatrix} 1 & x & z \\ 0 & 1 & y \\ 0 & 0 & 1 \end{pmatrix} \quad \text{with} \quad x, y, z \in \mathbb{R} .$$

For a fixed integer $k \neq 0$ consider the subset $\Gamma_k \subset Nil^3$ of matrices of the form

$$\begin{pmatrix} 1 & l & \frac{n}{k} \\ 0 & 1 & m \\ 0 & 0 & 1 \end{pmatrix} \quad \text{with} \quad l, m, n \in \mathbb{Z} .$$

Then Γ_k is a lattice in Nil^3, and it is known that every lattice in Nil^3 is isomorphic to precisely one Γ_k with $k > 0$; see Example 9.27 below.

9.2.1 Nomizu's Theorem

We will need some information about the cohomological structure of nilmanifolds.

For any Lie group G denote by $\Omega^G(G)$ the finite-dimensional vector space of left-invariant differential forms on G. We can identify this space with the space $\Lambda^* \mathfrak{g}^*$ of forms on the Lie algebra $\mathfrak{g}$. Since the differential commutes with pullback under smooth maps, the differential of a left-invariant form is left-invariant, hence it induces a differential

$$d \colon \Omega^G(G) \longrightarrow \Omega^G(G)$$

and thus a differential on $\Lambda^* \mathfrak{g}^*$ giving rise to a complex. The cohomology $H^*(\mathfrak{g}; \mathbb{R})$ of this complex is called the Lie algebra cohomology of $\mathfrak{g}$.

Assume now that G is simply connected and nilpotent, $\Gamma \subset G$ is a lattice and $M = \Gamma \backslash G$ is the associated nilmanifold. Since Γ acts on the left on G, the projection $\pi \colon G \to M$ induces a push-forward homomorphism

$$\pi_* \colon \Omega^G(G) \longrightarrow \Omega^*(M) \,,$$

which then induces a homomorphism

$$\pi_* \colon H^*(\mathfrak{g}; \mathbb{R}) \longrightarrow H^*_{dR}(M) \,.$$

The following theorem is very handy, for instance for calculating the Betti numbers of nilmanifolds.

Theorem 9.18 (Nomizu [Nom-54]) *The homomorphism*

$$\pi_* \colon H^*(\mathfrak{g}; \mathbb{R}) \longrightarrow H^*_{dR}(M)$$

is an isomorphism. In particular, the real cohomology of M is independent of the lattice Γ. Furthermore, the Euler characteristics of both $\mathfrak{g}$ and M vanish.

Remark 9.19 Nilmanifolds are parallelisable, because a basis of left-invariant vector fields on G descends to a global frame for TM. This implies that the Euler characteristic of M vanishes.

We consider two examples illustrating Nomizu's Theorem. Notice that if $e_1, \ldots, e_n$ is a basis for a Lie algebra with dual basis $\alpha_1, \ldots, \alpha_n$ of left-invariant one-forms, then the formula

$$d\alpha_i(e_j, e_k) = -\alpha_i([e_j, e_k]) \tag{9.2}$$

determines the two-forms $d\alpha_i$. This formula also implies that knowledge of the differentials $d\alpha_i$ is equivalent to knowledge of the commutators $[e_j, e_k]$.

Example 9.20   Consider the nilpotent Lie algebra $\mathfrak{nil}_3 = \mathfrak{n}_3$ as in Example 9.9 with simply connected nilpotent Lie group $Nil^3 = N_3$. This Lie algebra is three-dimensional and has a basis e_1, e_2, e_3 with only one non-zero commutator

$$[e_1, e_2] = e_3 .$$

Let $\alpha_1, \alpha_2, \alpha_3$ be the dual basis of one-forms. The formula (9.2) shows that

$$d\alpha_1 = d\alpha_2 = 0 ,$$
$$d\alpha_3 = \alpha_2 \wedge \alpha_1 .$$

Then $H^1(\mathfrak{g}; \mathbb{R})$ is spanned by the classes $[\alpha_1]$ and $[\alpha_2]$, and $H^2(\mathfrak{g}; \mathbb{R})$ is spanned by the classes $[\alpha_1 \wedge \alpha_3]$ and $[\alpha_2 \wedge \alpha_3]$. The form $\alpha_1 \wedge \alpha_2$ is closed, but also exact. It follows that any nilmanifold $M^3 = \Gamma \backslash Nil^3$ has Betti numbers

$$b_0(M) = b_3(M) = 1 ,$$
$$b_1(M) = b_2(M) = 2 .$$

Example 9.21   Consider the nilpotent Lie algebra $\mathfrak{nil}_4$ having a basis e_1, e_2, e_3 and e_4 for which the only non-zero commutators are

$$[e_1, e_2] = e_3 , \quad [e_1, e_3] = e_4 .$$

We have

$$d\alpha_1 = d\alpha_2 = 0 ,$$
$$d\alpha_3 = \alpha_2 \wedge \alpha_1 ,$$
$$d\alpha_4 = \alpha_3 \wedge \alpha_1 .$$

Then $H^1(\mathfrak{g}; \mathbb{R})$ is spanned by the classes $[\alpha_1]$ and $[\alpha_2]$, $H^2(\mathfrak{g}; \mathbb{R})$ is spanned by $[\alpha_1 \wedge \alpha_4]$ and $[\alpha_2 \wedge \alpha_3]$, and $H^3(\mathfrak{g}; \mathbb{R})$ is spanned by $[\alpha_1 \wedge \alpha_3 \wedge \alpha_4]$ and $[\alpha_2 \wedge \alpha_3 \wedge \alpha_4]$.

Let Nil^4 denote the simply connected nilpotent Lie group associated to $\mathfrak{nil}_4$. It follows that any nilmanifold $M^4 = \Gamma \backslash Nil^4$ has Betti numbers

$$b_0(M) = b_4(M) = 1 ,$$
$$b_1(M) = b_2(M) = b_3(M) = 2 .$$

This agrees with the fact that the Euler characteristic of M has to vanish.

9.2.2 Nilmanifolds as Towers of Circle Bundles

We now want to give a topological description of nilmanifolds.

Definition 9.22 We say that an n-dimensional smooth manifold M is the *total space of a tower of principal S^1-bundles* if there is a sequence of manifolds

$$M_0 = \{*\}, \quad M_1 = S^1, \quad M_2, \quad \ldots, \quad M_n = M$$

such that for each $i \geq 0$ the manifold M_{i+1} is the total space of a principal S^1-bundle over M_i:

$$S^1 \longrightarrow M_{i+1}$$
$$\downarrow$$
$$M_i \, .$$

This definition leads to the following characterisation of nilmanifolds.

Theorem 9.23 ([GOV-97]) *Nilmanifolds are up to diffeomorphism precisely the total spaces of towers of principal S^1-bundles.*

For example, the n-dimensional torus T^n is a tower of (trivial) principal S^1-bundles, corresponding to the nilmanifold associated to the abelian Lie group $\mathbb{R}^n$.

Corollary 9.24 (i) *The n-dimensional nilmanifolds, with $n \geq 1$, are precisely the principal S^1-bundles over $(n-1)$-dimensional nilmanifolds.*

(ii) *If M is an n-dimensional nilmanifold, with $n \geq 2$, then M is a (possibly non-principal) T^2-bundle over an $(n-2)$-dimensional nilmanifold.*

(iii) *If M is an n-dimensional nilmanifold, with $n \geq 1$, then M is a mapping torus $N \to M \to S^1$ whose general fibre N is an $(n-1)$-dimensional nilmanifold.*

Proof The first statement is clear from Theorem 9.23. For the second statement, consider the principal S^1-bundles

$$S^1 \longrightarrow M = M_n \xrightarrow{\pi_n} M_{n-1},$$
$$S^1 \longrightarrow M_{n-1} \xrightarrow{\pi_{n-1}} M_{n-2}.$$

The manifold M is closed and the composition $\pi = \pi_{n-1} \circ \pi_n$ is a surjective submersion, hence by Ehresmann's Fibration Theorem $\pi \colon M \to M_{n-2}$ is a smooth fibre bundle. The fibres

$$\pi^{-1}(p) = \pi_n^{-1}(\pi_{n-1}^{-1}(p)), \quad p \in M_{n-2},$$

are principal S^1-bundles over S^1, hence diffeomorphic to T^2.

The third claim follows similarly. $\qquad\qquad\square$

One can say more about the structure of the principal bundles in the theorem if one restricts to special classes of nilmanifolds. The two-step nilpotent Lie algebras are the simplest non-abelian Lie algebras. Recall from Definition 9.3 that a Lie algebra $\mathfrak{g}$ is two-step nilpotent if and only if

$$[\mathfrak{g}, [\mathfrak{g}, \mathfrak{g}]] = 0 \quad \text{and} \quad [\mathfrak{g}, \mathfrak{g}] \neq 0 \,,$$

i.e. $[\mathfrak{g}, \mathfrak{g}]$ is a non-zero subset of the center of $\mathfrak{g}$.

We can explicitly construct two-step nilpotent Lie algebras as follows. Let $\mathfrak{z}$ and $\mathfrak{v}$ denote real vector spaces of dimension m and n respectively, with bases $\{Z_1, ..., Z_m\}$ and $\{X_1, ..., X_n\}$, respectively. Let

$$\left[Z_i, Z_j\right] = 0 \,,$$
$$\left[X_i, Z_j\right] = 0 \,,$$
$$\left[X_i, X_j\right] = \sum_k c_{ij}^k Z_k \,,$$

where c_{ij}^k are constants that are antisymmetric in the indices i and j. This defines a Lie algebra structure on

$$\mathfrak{g} = \mathfrak{z} \oplus \mathfrak{v} \,,$$

which is two-step nilpotent if at least one of the structure constants c_{ij}^k is non-zero. It is clear that every two-step nilpotent Lie algebra can be obtained in this way. If the structure constants c_{ij}^k are rational, then the associated simply connected nilpotent Lie group G has a lattice.

The two-step nilpotent Lie algebras are interesting for us because of the following special case of Theorem 9.23.

Theorem 9.25 (Palais–Stewart [PS-61])   *The nilmanifolds associated to simply connected r-step nilpotent Lie groups, with $r \leq 2$, are up to diffeomorphism precisely the principal T^m-bundles over T^n.*

Remark 9.26   The case $r = 0$ is trivial and the nilmanifolds associated to one-step nilpotent (abelian) Lie groups are tori T^{m+n}. The nilmanifolds $M = \Gamma \backslash G$ associated to simply connected two-step nilpotent Lie groups G are precisely the non-trivial principal T^m-bundles over T^n, because Γ is isomorphic to $\pi_1(M)$ and non-abelian.

Example 9.27   We discuss the structure of three-dimensional nilmanifolds. In dimension 3 there are two simply connected nilpotent Lie groups: the abelian Lie group $\mathbb{R}^3$ and the two-step nilpotent Lie group Nil^3.

Nilmanifolds of $\mathbb{R}^3$ are tori T^3 and the nilmanifolds M associated to Nil^3 are precisely the non-trivial principal S^1-bundles over T^2. If the Euler class of this

S^1-bundle is k times the generator of $H^2(T^2; \mathbb{Z})$, with an integer $k \in \mathbb{Z} \setminus \{0\}$, then the Gysin sequence shows that

$$H_0(M; \mathbb{Z}) = H_3(M; \mathbb{Z}) \cong \mathbb{Z},$$
$$H_1(M; \mathbb{Z}) \cong \mathbb{Z}^2 \oplus \mathbb{Z}_k, \quad H_2(M; \mathbb{Z}) \cong \mathbb{Z}^2.$$

This agrees with the calculation in Example 9.20. By Corollary 9.24 the three-dimensional nilmanifolds also have the structure of mapping tori over S^1 with fibre T^2.

Example 9.28 We can similarly discuss the structure of four-dimensional nilmanifolds. In this dimension there are three simply connected nilpotent Lie groups: the abelian Lie group $\mathbb{R}^4$, the two-step nilpotent Lie group $Nil^3 \times \mathbb{R}$ and the three-step nilpotent Lie group Nil^4. By Corollary 9.24 the four-dimensional nilmanifolds are T^2-bundles over T^2 and also mapping tori over S^1 with fibre a three-dimensional nilmanifold.

The nilmanifolds associated to $\mathbb{R}^4$ are tori T^4 and the nilmanifolds associated to $Nil^3 \times \mathbb{R}$ are precisely the non-trivial principal T^2-bundles over T^2. The nilmanifolds associated to Nil^4 are non-trivial, non-principal T^2-bundles over T^2.

A calculation using the Gysin sequence shows that Nil^4-manifolds M, which are characterised among four-dimensional nilmanifolds by $b_1(M) = 2$, are the principal S^1-bundles with non-torsion Euler class over Nil^3-manifolds.

9.2.3 Infra-nilmanifolds

There is a class of manifolds called infra-nilmanifolds that is a small but important generalisation of the class of nilmanifolds. So far, for a Lie group G we considered only the action of G on itself by left translations, and we looked for co-compact discrete subgroups. One can also consider the action of $\mathrm{Aut}(G)$, the Lie group of smooth group automorphisms of G, on G.

Definition 9.29 The *affine group* of G, denoted by $\mathrm{Aff}(G)$, is the semidirect product $G \rtimes \mathrm{Aut}(G)$.

An element $(g, \varphi) \in \mathrm{Aff}(G)$ acts on G by

$$(g, \varphi)(x) = g \cdot \varphi(x) = (l_g \circ \varphi)(x),$$

where l_g is the action of g on G by left translation.

Definition 9.30 An *infra-nilmanifold* is a quotient $\Gamma \backslash G$ of a simply connected nilpotent Lie group G, where $\Gamma \subset G \rtimes K \subset \mathrm{Aff}(G)$ is a torsion-free discrete cocompact subgroup of $\mathrm{Aff}(G)$ contained in $G \rtimes K$ for some compact subgroup $K \subset \mathrm{Aut}(G)$.

The assumptions imply that $\Gamma \cap G$ is a uniform lattice in G and is of finite index in Γ. Therefore every infra-nilmanifold admits a finite covering by a nilmanifold.

Example 9.31 If G is abelian, then $G = \mathbb{R}^n$ and the only nilmanifold is the torus T^n. However, since $O(n)$ is maximal compact in $\mathrm{Aut}(\mathbb{R}^n) = GL_n(\mathbb{R})$, the infra-nilmanifolds in this case are precisely the closed flat Riemannian n-manifolds. It makes sense to call them infra-abelian.

Given that the number of closed flat Riemannian n-manifolds grows very quickly with n, we see that there are many more infra-nilmanifolds than there are nilmanifolds.

Borrowing additional terminology from this crucial example, one may call $\Gamma/(\Gamma \cap G)$ the holonomy group of the infra-nilmanifold $\Gamma \backslash G$. It is finite and is the Galois group of the covering of the infra-nilmanifold by its associated nilmanifold.

9.3 Lagrangian Foliations and Künneth Structures

For a Lie group G we now want to consider left-invariant geometric structures, i.e. structures invariant under all left translations l_g of G. Such left-invariant structures on G are completely determined by the corresponding linear structures on the Lie algebra $\mathfrak{g}$. We will thus speak (algebraically) of foliations, of symplectic structures, of Künneth structures or hypersymplectic structures on the Lie algebra $\mathfrak{g}$, by which we really mean (geometrically) the corresponding left-invariant structures on the Lie group G.

For instance, a foliation on the Lie algebra $\mathfrak{g}$ is given, according to the Frobenius Theorem, by a Lie subalgebra $\mathfrak{h} \subset \mathfrak{g}$. A symplectic structure on $\mathfrak{g}$ is a closed, non-degenerate two-form $\omega \in \Lambda^2 \mathfrak{g}^*$. It is then clear how to define the notions of Lagrangian foliations, of Künneth structures and of hypersymplectic structures on $\mathfrak{g}$.

Assume that G is a simply connected nilpotent Lie group and $\Gamma \subset G$ is a lattice. Then every left-invariant structure on G is invariant under the left action of Γ on G and thus descends to a well-defined structure on the nilmanifold $M = \Gamma \backslash G$. Notice that distributions (foliations) on M defined by vector subspaces (Lie subalgebras) $\mathfrak{h} \subset \mathfrak{g}$ are trivial as vector subbundles of TM; see Remark 9.19.

Since the structures we are interested in can often be defined quite easily on specific nilpotent Lie algebras, this method yields numerous examples of the corresponding structures on the associated nilmanifolds.

Example 9.32 Let $\mathfrak{g}$ be a non-abelian nilpotent Lie algebra. Consider

$$\mathfrak{h} = \mathfrak{g}_{(2)} = [\mathfrak{g}, \mathfrak{g}] .$$

Then $0 \neq \mathfrak{h} \neq \mathfrak{g}$ is a non-trivial ideal and hence a non-trivial subalgebra of $\mathfrak{g}$. The foliation $\mathfrak{h}$ on $\mathfrak{g}$ defines a non-trivial foliation on any nilmanifold of the form $M = \Gamma\backslash G$.

Concerning symplectic structures on nilmanifolds in general we note the following.

Proposition 9.33 *Let ω' be an arbitrary symplectic form on a nilmanifold $M = \Gamma\backslash G$. Then there exists a left-invariant symplectic form ω on G, such that the induced symplectic form $\pi_*\omega$ on M represents the same de Rham cohomology class as ω'.*

Proof Suppose ω' is some symplectic form on the nilmanifold $M^{2n} = \Gamma\backslash G$. By Nomizu's Theorem (Theorem 9.18) we can find in the class $[\omega'] \in H^2_{dR}(M)$ a representative that comes from a left-invariant closed two-form ω on G. We have $[\omega]^n \neq 0$, so ω is non-degenerate because it is left-invariant. $\square$

This implies the following.

Corollary 9.34 *A nilmanifold $M = \Gamma\backslash G$ admits a symplectic form if and only if the Lie algebra $\mathfrak{g}$ admits a symplectic form.*

Concerning the existence of Lagrangian foliations, there is the following general theorem due to Baues and Cortés; see Corollary 3.13 in [BC-16]).

Theorem 9.35 (Baues–Cortés [BC-16]) *Every symplectic structure on a nilpotent Lie algebra admits a Lagrangian foliation.*

Together with Corollary 9.34, this implies the following statement for nilmanifolds.

Corollary 9.36 *Let (M, ω') be a symplectic nilmanifold. Then there exists a symplectic form ω on M representing the same de Rham cohomology class $[\omega] = [\omega']$ such that ω admits a Lagrangian foliation.*

We now want to discuss some low-dimensional examples explicitly, without using Theorem 9.35. This discussion will eventually yield the conclusion that there is no analogue of this theorem for Künneth structures.

9.4 Low-dimensional Examples

In this section we discuss explicit examples of Künneth structures on nilmanifolds up to dimension 6.

There is nothing to prove in the two-dimensional case, because the abelian Lie algebra $\mathbb{R}^2$ is the unique nilpotent two-dimensional Lie algebra. For a symplectic form ω on $\mathbb{R}^2$, the Linear Darboux Theorem (Theorem 2.5) implies that there exists a basis e_1, e_2 with dual one-forms α_1, α_2 so that

$$\omega = \alpha_1 \wedge \alpha_2 \,.$$

A Künneth structure is given by

$$\mathfrak{f} = \mathrm{span}\{e_1\} \,,$$
$$\mathfrak{h} = \mathrm{span}\{e_2\} \,.$$

In dimension 4 there are up to isomorphism three nilpotent Lie algebras: the abelian Lie algebra $\mathbb{R}^4$ and the Lie algebras $\mathfrak{nil}_3 \oplus \mathbb{R}$ and $\mathfrak{nil}_4$. All of them admit symplectic forms. For the abelian Lie algebra $\mathbb{R}^4$ the Linear Darboux Theorem again implies that every symplectic form ω admits a Künneth structure.

9.4.1 The Lie Group $Nil^3 \times \mathbb{R}$

We now consider the nilpotent Lie algebra $\mathfrak{nil}_3 \oplus \mathbb{R}$ with basis e_1, e_2, e_3, e_4 and non-zero commutator

$$[e_1, e_2] = e_3 \,.$$

Equivalently we have for the dual one-forms

$$d\alpha_1 = d\alpha_2 = d\alpha_4 = 0 \,,$$
$$d\alpha_3 = \alpha_2 \wedge \alpha_1 \,.$$

We want to consider symplectic forms ω on $\mathfrak{nil}_3 \oplus \mathbb{R}$ and make the general ansatz

$$\omega = \sum_{i<j} A_{ij} \alpha_i \wedge \alpha_j \,.$$

Then

$$d\omega = A_{34} \alpha_2 \wedge \alpha_1 \wedge \alpha_4 \,,$$

hence $A_{34} = 0$ is necessary for ω to be symplectic. It follows that

$$\omega = A_{12}\alpha_1 \wedge \alpha_2 + A_{13}\alpha_1 \wedge \alpha_3 + A_{14}\alpha_1 \wedge \alpha_4 + A_{23}\alpha_2 \wedge \alpha_3 + A_{24}\alpha_2 \wedge \alpha_4$$

and
$$\omega^2 = 2(-A_{13}A_{24} + A_{14}A_{23})\alpha_1 \wedge \alpha_2 \wedge \alpha_3 \wedge \alpha_4 \, ,$$
hence we need $A_{13}A_{24} \neq A_{14}A_{23}$.

We consider three cases and in each case exhibit a Künneth structure made up of complementary Lagrangian subalgebras $\mathfrak{f}$, $\mathfrak{h}$:

- $A_{13} \neq 0$:

$$\mathfrak{f} = \text{span}\{e_1, e_4 - (A_{14}/A_{13})e_3\} \, ,$$
$$\mathfrak{h} = \text{span}\{e_2 - (A_{23}/A_{13})e_1, e_3\} \, ,$$

- $A_{24} \neq 0$:

$$\mathfrak{f} = \text{span}\{e_1 - (A_{14}/A_{24})e_2, e_4\} \, ,$$
$$\mathfrak{h} = \text{span}\{e_2, e_3 - (A_{23}/A_{24})e_4\} \, ,$$

- $A_{13} = A_{24} = 0$:

$$\mathfrak{f} = \text{span}\{e_1, e_3\} \, ,$$
$$\mathfrak{h} = \text{span}\{e_2, e_4\} \, .$$

We have proved the following strengthening of Theorem 9.35 in the case of $\mathfrak{nil}_3 \oplus \mathbb{R}$.

Proposition 9.37 *Every symplectic form on $\mathfrak{nil}_3 \oplus \mathbb{R}$ admits a Künneth structure.*

In particular, with Example 9.28 and the discussion of Künneth structures on the abelian Lie algebra $\mathbb{R}^4$, we obtain the following.

Corollary 9.38 *Let M be a principal T^2-bundle over T^2 with an arbitrary symplectic form ω'. Then there exists a symplectic form ω on M representing the same de Rham cohomology class $[\omega] = [\omega']$ such that ω admits a Künneth structure.*

9.4.2 The Lie Group Nil^4

The nilpotent Lie algebra $\mathfrak{nil}_4$ admits bases e_1, e_2, e_3, e_4 with non-zero commutators given by

$$[e_1, e_2] = e_3 \, ,$$
$$[e_1, e_3] = e_4 \, .$$

We will say that such a basis is in standard form.

Lemma 9.39 *Let ω be a symplectic form on the Lie algebra $\mathfrak{nil}_4$. Then there exists a basis e_1, e_2, e_3, e_4 in standard form such that the symplectic form is given in the dual basis by*

$$\omega = \alpha_1 \wedge \alpha_4 + \alpha_2 \wedge \alpha_3 .$$

Proof Let e_1, e_2, e_3 and e_4 be a basis for $\mathfrak{nil}_4$ in standard form. Then

$$d\alpha_1 = d\alpha_2 = 0 ,$$
$$d\alpha_3 = \alpha_2 \wedge \alpha_1 ,$$
$$d\alpha_4 = \alpha_3 \wedge \alpha_1 .$$

We can write any given symplectic form ω as

$$\omega = \sum_{i<j} A_{ij} \alpha_i \wedge \alpha_j$$

for suitable coefficients A_{ij}. Then

$$0 = d\omega = -A_{24}\alpha_2 \wedge \alpha_3 \wedge \alpha_1 + A_{34}\alpha_2 \wedge \alpha_1 \wedge \alpha_4 .$$

Hence $A_{24} = A_{34} = 0$. We can then write ω as

$$\omega = \alpha_1 \wedge (A_{12}\alpha_2 + A_{13}\alpha_3 + A_{14}\alpha_4) + A_{23}\alpha_2 \wedge \alpha_3 .$$

The form ω is non-degenerate if and only if $A_{14}A_{23} \neq 0$. We introduce a non-zero constant

$$C = \sqrt[5]{\frac{A_{23}}{A_{14}^2}} .$$

Let

$$\alpha_1' = \frac{1}{C}\alpha_1 ,$$
$$\alpha_2' = C^3 A_{14}\alpha_2 ,$$
$$\alpha_3' = C^2 (A_{13}\alpha_2 + A_{14}\alpha_3) ,$$
$$\alpha_4' = C(A_{12}\alpha_2 + A_{13}\alpha_3 + A_{14}\alpha_4) .$$

Then α_1', α_2', α_3', and α_4' are a basis of one-forms. We have

$$\omega = \alpha_1' \wedge \alpha_4' + \alpha_2' \wedge \alpha_3'$$

and

$$d\alpha_1' = d\alpha_2' = 0 ,$$
$$d\alpha_3' = \alpha_2' \wedge \alpha_1' ,$$
$$d\alpha_4' = \alpha_3' \wedge \alpha_1' .$$

Therefore the dual basis of $\mathfrak{nil}_4$ has non-zero commutators

$$[e_1', e_2'] = e_3' \,,$$
$$[e_1', e_3'] = e_4' \,.$$

It follows that e_1', e_2', e_3', e_4' form a basis in standard form. $\qquad\qquad\square$

We can now prove the following.

Proposition 9.40 *Consider the Lie algebra* $\mathfrak{nil}_4$.

(i) *Every symplectic structure on* $\mathfrak{nil}_4$ *admits a Lagrangian foliation.*
(ii) *There is no symplectic structure on* $\mathfrak{nil}_4$ *that admits a Künneth structure.*

Proof Let ω be a symplectic structure on $\mathfrak{nil}_4$. According to Lemma 9.39 we can choose a basis e_1, e_2, e_3, e_4 in standard form with non-zero commutators

$$[e_1, e_2] = e_3 \,,$$
$$[e_1, e_3] = e_4 \,,$$

so that

$$\omega = \alpha_1 \wedge \alpha_4 + \alpha_2 \wedge \alpha_3 \,.$$

For the first statement, note that the span of $\{e_2, e_4\}$ and similarly the span of $\{e_3, e_4\}$ are Lagrangian foliations on $\mathfrak{nil}_4$.

To prove the non-existence of Künneth structures, suppose that $\mathfrak{nil}_4$ has two complementary two-dimensional foliations $\mathfrak{f}$ and $\mathfrak{h}$. Then one of the subalgebras, say $\mathfrak{f}$, has a basis vector of the form

$$f_1 = e_1 + \sum_{i=2}^{4} a_i e_i$$

with real coefficients a_i. Any other basis vector of $\mathfrak{f}$ can be assumed to be of the form

$$f_2 = \sum_{j=2}^{4} b_j e_j \,.$$

According to Proposition 9.11 the Lie subalgebra $\mathfrak{f}$ is nilpotent, hence abelian, since it is two-dimensional. Thus

$$[f_1, f_2] = b_2 e_3 + b_3 e_4$$

implies that $b_2 = b_3 = 0$ and $\mathfrak{f}$ is spanned by a basis of the form

$$f_1' = e_1 + a_2 e_2 + a_3 e_3 \,,$$
$$f_2' = e_4 \,.$$

However,

$$\omega(f_1', f_2') = 1$$

implies that $\mathfrak{f}$ cannot be Lagrangian. ☐

This proposition shows that Künneth structures do not always exist on symplectic nilpotent Lie algebras. In particular, Theorem 9.35 has no analogue for Künneth structures.

If a nilpotent Lie algebra $\mathfrak{g}$ does not admit a Künneth structure, an associated nilmanifold of the form $M = \Gamma \backslash G$ could still have a Künneth structure. Such a Künneth structure would come from a Künneth structure on G that would be left-invariant under the action of the lattice Γ, but not under the full group G.

9.4.3 Dimension 6

We now move on to the six-dimensional case, where for the first time one has a large number of symplectic nilpotent Lie algebras.

Up to isomorphism there are 26 nilpotent six-dimensional Lie algebras which admit a symplectic form. They were classified from different points of view by Salamon [Sal-01], by Khakimdjanov, Goze and Medina [KGM-04], and by Bazzoni and Muñoz [BM-12]. Hamilton [Ham-19] clarified the comparison of these classifications, and determined which of the Lie algebras admit a Künneth structure. Without going into the details, we give an overview of the results in Table 9.1. In this table, the columns with headers *I*, *II* and *III* use the notations of [Sal-01], [KGM-04] and [BM-12], respectively. For example, in the notation of [BM-12], A_n is the abelian Lie algebra $\mathbb{R}^n$, $L_3 = \mathfrak{nil}_3$ and $L_4 = \mathfrak{nil}_4$.

Table 9.1 also gives the values of the Betti numbers b_1 and b_2 for the corresponding nilmanifolds. Since the Betti numbers satisfy

$$2 - 2b_1 + 2b_2 - b_3 = 0,$$

because the Euler characteristic vanishes according to Nomizu's Theorem (Theorem 9.18), the values of b_1 and b_2 determine b_3.

For wedge products of one-forms we use the abbreviation

$$\alpha_{ij} = \alpha_i \wedge \alpha_j.$$

The differentials $d\alpha_1, \ldots, d\alpha_6$ (related to the commutators of the dual basis $e_1, \ldots, e_6$ by equation (9.2)) are given in the columns for the structure constants. In all cases α_1 and α_2 are closed.

The entry *not Künneth* in the table means that no symplectic structure on

Algebraic structure					Structure constants				Symplectic form	Künneth structure
b_1	b_2	I	II	III	$d\alpha_3$	$d\alpha_4$	$d\alpha_5$	$d\alpha_6$		
6	15	(0,0,0,0,0,0)	26	A_6	0	0	0	0	$\alpha_{12}+\alpha_{34}+\alpha_{56}$	$\{e_1,e_3,e_5\},\{e_2,e_4,e_6\}$
5	11	(0,0,0,0,0,12)	25	$L_3\oplus A_3$	0	0	0	α_{12}	$\alpha_{16}+\alpha_{23}+\alpha_{45}$	$\{e_1,e_3,e_4\},\{e_2,e_5,e_6\}$
4	9	(0,0,0,0,12,13)	23	$L_{5,2}\oplus A_1$	0	0	α_{12}	α_{13}	$\alpha_{16}+\alpha_{24}+\alpha_{36}$	$\{e_1,e_4,e_6\},\{e_2,e_3,e_5\}$
4	8	(0,0,0,0,12,34)	24	$L_3\oplus L_3$	0	0	α_{12}	α_{34}	$\alpha_{15}+\alpha_{24}+\alpha_{36}$	$\{e_1,e_4,e_6\},\{e_2,e_3,e_5\}$
4	8	(0,0,0,0,12,14+23)	17	$L_{6,1}$	0	0	α_{12}	$\alpha_{13}+\alpha_{24}$	$\alpha_{16}+\alpha_{23}-\alpha_{45}$	$\{e_1,e_2,e_5\},\{e_3,e_4,e_6\}$
4	8	(0,0,0,13+42,14+23)	16	$L_{6,2}$	0	0	$\alpha_{13}-\alpha_{24}$	$\alpha_{14}+\alpha_{23}$	$\alpha_{16}+\alpha_{25}+\alpha_{34}$	$\{e_1,e_3,e_5\},\{e_1+e_2,e_4,e_5-e_6\}$
4	7	(0,0,0,0,12,15)	22	$L_4\oplus A_2$	0	0	α_{12}	α_{15}	$\alpha_{16}+\alpha_{25}+\alpha_{34}$	not Künneth
4	7	(0,0,0,0,12,14+25)	21	$L_{5,3}\oplus A_1$	0	0	α_{12}	$\alpha_{15}+\alpha_{23}$	$\alpha_{16}+\alpha_{25}-\alpha_{35}$	$\{e_1,e_3,e_4\},\{e_2,e_5,e_6\}$
3	8	(0,0,0,12,13,23)	18	$L_{6,4}$	0	α_{12}	α_{13}	α_{23}	$\alpha_{16}+2\alpha_{25}+\alpha_{34}$	$\{e_1,e_2,e_4\},\{e_3,e_5,e_6\}$
3	6	(0,0,0,12,13,14)	14	$L_{6,5}$	0	α_{12}	α_{13}	α_{14}	$\alpha_{16}+\alpha_{25}+\alpha_{34}$	$\{e_1,e_3,e_5\},\{e_2,e_4,e_6\}$
3	6	(0,0,0,12,13,24)	15	$L_{6,6}$	0	α_{12}	α_{13}	α_{24}	$\alpha_{15}+\alpha_{25}-\alpha_{26}+\alpha_{34}$	$\{e_1-e_2,e_3,e_5\},\{e_1+e_5,e_4,e_6\}$
3	6	(0,0,0,12,13,14+23)	13	$L_{6,9}$	0	α_{12}	α_{13}	$\alpha_{14}+\alpha_{23}$	$\alpha_{16}+2\alpha_{25}-\alpha_{34}$	$\{e_1,e_3,e_5\},\{e_2,e_4,e_6\}$
3	5	(0,0,0,12,14,13+42)	11	$L_{6,10}$	0	α_{12}	α_{14}	$\alpha_{23}+\alpha_{24}$	$\alpha_{16}+\alpha_{25}-\alpha_{34}$	$\{e_1,e_4,e_5\},\{e_2-e_4,e_3+e_5,e_6\}$
3	5	(0,0,0,12,14,13+42)	10	$L_{6,11}$	0	α_{12}	α_{14}	$\alpha_{13}+\alpha_{24}$	$\alpha_{16}+\alpha_{25}-\alpha_{26}-\alpha_{34}$	$\{e_1,e_4,e_5\},\{e_1+e_2-e_4,e_3+e_5,e_6\}$
3	5	(0,0,0,12,13+42,14+23)	12	$L_{6,12}$	0	α_{12}	$\alpha_{14}+\alpha_{23}$	$\alpha_{13}-\alpha_{24}$	$-\alpha_{15}+6\alpha_{26}+7\alpha_{34}$	$\{e_2,e_3,e_5\},\{e_2-2e_1,e_3-e_4,-3e_5+e_6\}$
3	5	(0,0,0,12,14,15)	19	$L_{5,4}\oplus A_1$	0	α_{12}	α_{14}	α_{15}	$\alpha_{13}+\alpha_{26}-\alpha_{45}$	$\{e_1,e_5,e_6\},\{e_2,e_3,e_4\}$
3	5	(0,0,0,12,14,15+23)	9	$L_{6,13}$	0	α_{12}	α_{14}	$\alpha_{15}+\alpha_{23}$	$\alpha_{13}+\alpha_{26}-\alpha_{45}$	not Künneth
3	5	(0,0,0,12,14,15+23+24)	20	$L_{5,6}\oplus A_1$	0	α_{12}	α_{14}	$\alpha_{15}+\alpha_{23}+\alpha_{24}$	$\alpha_{13}+\alpha_{26}-\alpha_{45}$	not Künneth
3	5	(0,0,0,12,14-23,15+24)	7	$L_{6,14}$	0	α_{12}	α_{14}	$\alpha_{15}+\alpha_{23}+\alpha_{24}$	$\alpha_{13}+\alpha_{26}-\alpha_{45}$	not Künneth
3	4	(0,0,0,12,14-23,15+34)	8	$L_{6,15}$	0	α_{12}	α_{14}	$\alpha_{15}-\alpha_{34}$	$\alpha_{16}+\alpha_{24}-\alpha_{35}$	not Künneth
2	4	(0,0,12,13,23,14)	6	$L_{6,16}$	α_{12}	α_{13}	α_{23}	α_{14}	$\alpha_{15}+\alpha_{24}+\alpha_{26}-\alpha_{34}$	$\{e_1,e_4,e_6\},\{e_2,e_3,e_5\}$
2	4	(0,0,12,13,23,14+25)	4	$L_{6,17}^{+}$	α_{12}	α_{13}	α_{23}	$\alpha_{14}+\alpha_{25}$	$\alpha_{16}+\alpha_{15}+\alpha_{24}+\alpha_{35}$	not Künneth
2	4	(0,0,12,13,23,14-25)	5	$L_{6,17}^{-}$	α_{12}	α_{13}	α_{23}	$\alpha_{14}-\alpha_{25}$	$\alpha_{15}-\alpha_{16}+\alpha_{24}+\alpha_{35}$	not Künneth
2	3	(0,0,12,13,14,15)	3	$L_{6,18}$	α_{12}	α_{13}	α_{14}	α_{15}	$\alpha_{16}+\alpha_{25}-\alpha_{34}$	not Künneth
2	3	(0,0,12,13,14,23+15)	2	$L_{6,19}$	α_{12}	α_{13}	α_{14}	$\alpha_{15}+\alpha_{23}$	$\alpha_{16}+\alpha_{24}+\alpha_{25}-\alpha_{34}$	not Künneth
2	3	(0,0,12,13,14+23,24+15)	1	$L_{6,21}$	α_{12}	α_{13}	$\alpha_{14}+\alpha_{23}$	$\alpha_{15}+\alpha_{24}$	$2\alpha_{16}+\alpha_{25}+\alpha_{34}$	not Künneth

Table 9.1 Symplectic and Künneth structures on six-dimensional nilpotent Lie algebras. Columns b_1, b_2 indicate the Betti numbers, the algebraic structures I, II and III refer to the notation in [Sal-01], [KGM-04] and [BM-12], respectively. The structure constants (with $d\alpha_i = 0$ for $i = 1, 2$) are from [BM-12], the symplectic forms are from [KGM-04] and [BM-12].

that Lie algebra admits a Künneth structure. Out of the 26 symplectic nilpotent six-dimensional Lie algebras, 16 admit a Künneth structure and 10 do not.

Here we discuss only one of the entries in the table, to indicate how the discussion in [Ham-19] proceeds.

Example 9.41   We consider the Lie algebra $L_{6,1}$, which is not the direct sum of lower-dimensional Lie subalgebras. The structure constants are given by

$$d\alpha_1 = d\alpha_2 = d\alpha_3 = d\alpha_4 = 0 \,,$$
$$d\alpha_5 = \alpha_1 \wedge \alpha_2 \,,$$
$$d\alpha_6 = \alpha_1 \wedge \alpha_3 + \alpha_2 \wedge \alpha_4 \,.$$

This corresponds to the non-zero commutators of the dual basis

$$[e_1, e_2] = -e_5 \,,$$
$$[e_1, e_3] = -e_6 \,, \qquad (9.3)$$
$$[e_2, e_4] = -e_6 \,.$$

A symplectic form on $L_{6,1}$ is given by

$$\omega = \alpha_1 \wedge \alpha_6 + \alpha_2 \wedge \alpha_3 - \alpha_4 \wedge \alpha_5 \,.$$

Indeed,

$$\omega^3 = -6\alpha_1 \wedge \alpha_2 \wedge \alpha_3 \wedge \alpha_4 \wedge \alpha_5 \wedge \alpha_6 \neq 0$$

and

$$d\omega = -\alpha_1 \wedge \alpha_2 \wedge \alpha_4 + \alpha_4 \wedge \alpha_1 \wedge \alpha_2 = 0 \,.$$

A Künneth structure on $L_{6,1}$ for this symplectic form is given by

$$\mathfrak{f} = \text{span}\{e_1, e_2, e_5\} \,,$$
$$\mathfrak{h} = \text{span}\{e_3, e_4, e_6\} \,.$$

It is clear that these subspaces are Lagrangian and equations (9.3) show that they are closed under commutators.

9.5 Hypersymplectic Structures on Nilmanifolds

In this section we give some examples of hypersymplectic structures on nilpotent Lie algebras.

First of all, recall the case of $Nil^3 \times \mathbb{R}$, discussed in Subsection 9.4.1. The abundance of symplectic forms and of Künneth structures in that case suggests that there should be an underlying hypersymplectic structure. This is indeed the case. Recall that $Nil^3 \times \mathbb{R}$ has a framing by left-invariant one-forms $\alpha_1, \ldots, \alpha_4$ with $d\alpha_3 = \alpha_2 \wedge \alpha_1$, and α_i closed for $i \neq 3$. The three left-invariant two-forms,

$$\alpha = \alpha_3 \wedge \alpha_2 + \alpha_1 \wedge \alpha_4 \,,$$

$$\beta = \alpha_3 \wedge \alpha_1 - \alpha_2 \wedge \alpha_4 \,,$$

$$\omega = \alpha_3 \wedge \alpha_1 + \alpha_2 \wedge \alpha_4 \,,$$

are closed and non-degenerate, and therefore symplectic. Moreover, we can directly read off the recursion operators and see that they satisfy $A^2 = B^2 = -J^2 = \mathrm{Id}$. Therefore, the three forms define a left-invariant hypersymplectic structure fitting into the defining diagram

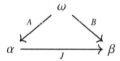

This structure descends to all nilmanifolds for $Nil^3 \times \mathbb{R}$. We will prove in Proposition 10.11 of Chapter 10 that these nilmanifolds account for all non-toral closed hypersymplectic four-manifolds.

Since hypersymplectic manifolds have dimensions divisible by 4, we now move on to dimension 8. The following construction, with a different notation, is part of Example 1 on page 261 in the paper of Fino, Pedersen, Poon and Sørensen [FPPS-04]. We consider the eight-dimensional two-step nilpotent Lie algebra $\mathfrak{g}$ with basis $e_1, e_2, \ldots, e_8$ and the only non-trivial commutators given by

$$[e_3, e_1] = e_7 \,,$$

$$[e_4, e_2] = e_7 \,,$$

$$[e_4, e_1] = e_8 \,,$$

$$[e_3, e_2] = -e_8 \,.$$

This implies for the dual basis $\alpha_1, \alpha_2, \ldots, \alpha_8$ of one-forms

$$d\alpha_i = 0, \quad \text{for} \quad i = 1, \ldots, 6 \,,$$

$$d\alpha_7 = \alpha_1 \wedge \alpha_3 + \alpha_2 \wedge \alpha_4 \,, \tag{9.4}$$

$$d\alpha_8 = \alpha_1 \wedge \alpha_4 - \alpha_2 \wedge \alpha_3 \,.$$

Mapping

$$\alpha_1 \mapsto \alpha_1' \,, \quad \alpha_2 \mapsto \alpha_2' \,, \quad \alpha_3 \mapsto \alpha_4' \,, \quad \alpha_4 \mapsto \alpha_3' \,,$$
$$\alpha_5 \mapsto \alpha_8' \,, \quad \alpha_6 \mapsto \alpha_7' \,, \quad \alpha_7 \mapsto \alpha_6' \,, \quad \alpha_8 \mapsto \alpha_5' \,,$$

it follows that $\mathfrak{g}$ is isomorphic to $L_{6,2} \oplus \mathbb{R}^2$ in the notation of Table 9.1.

We claim that the following three two-forms define a hypersymplectic structure on $\mathfrak{g}$:

$$\omega = \alpha_1 \wedge \alpha_8 - \alpha_2 \wedge \alpha_7 + \alpha_3 \wedge \alpha_6 - \alpha_4 \wedge \alpha_5 \,,$$

$$\alpha = \alpha_1 \wedge \alpha_6 + \alpha_2 \wedge \alpha_5 + \alpha_3 \wedge \alpha_8 + \alpha_4 \wedge \alpha_7 \,,$$

$$\beta = \alpha_1 \wedge \alpha_5 - \alpha_2 \wedge \alpha_6 + \alpha_3 \wedge \alpha_7 - \alpha_4 \wedge \alpha_8 \,.$$

To verify this, one first checks that ω, α, β are indeed symplectic. Next, one defines

$$Je_1 = e_2 \,, \qquad\qquad Je_2 = -e_1 \,,$$
$$Je_3 = e_4 \,, \qquad\qquad Je_4 = -e_3 \,,$$
$$Je_5 = e_6 \,, \qquad\qquad Je_6 = -e_5 \,,$$
$$Je_7 = e_8 \,, \qquad\qquad Je_8 = -e_7 \,,$$

$$Ae_1 = e_3 \,, \qquad\qquad Ae_3 = e_1 \,,$$
$$Ae_2 = -e_4 \,, \qquad\qquad Ae_4 = -e_2 \,,$$
$$Ae_5 = -e_7 \,, \qquad\qquad Ae_7 = -e_5 \,,$$
$$Ae_6 = e_8 \,, \qquad\qquad Ae_8 = e_6 \,,$$

and

$$Be_1 = -e_4 \,, \qquad\qquad Be_4 = -e_1 \,,$$
$$Be_2 = -e_3 \,, \qquad\qquad Be_3 = -e_2 \,,$$
$$Be_5 = e_8 \,, \qquad\qquad Be_8 = e_5 \,,$$
$$Be_6 = e_7 \,, \qquad\qquad Be_7 = e_6 \,.$$

It is clear that

$$J^2 = -\mathrm{Id}_\mathfrak{g} \,, \quad A^2 = B^2 = \mathrm{Id}_\mathfrak{g} \,,$$

and it is not difficult to check that

$$AJ = B$$

and

$$\alpha(X, Y) = \omega(AX, Y) \,,$$
$$\beta(X, Y) = \omega(BX, Y) \quad \forall X, Y \in \mathfrak{g} \,.$$

This means that J, A and B are indeed the recursion operators intertwining the three symplectic forms. It follows that we have the commutative diagram of recursion operators

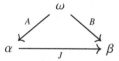

and ω, α, β indeed define a hypersymplectic structure on the eight-dimensional nilpotent Lie algebra $\mathfrak{g}$.

Remark 9.42   Since the hypersymplectic structure is left-invariant, it induces a hypersymplectic structure on every associated nilmanifold $M = \Gamma \backslash G$. By Theorem 9.25, the nilmanifolds M have the structure of principal T^4-bundles over T^4.

9.6 Anosov Symplectomorphisms

We can define Anosov automorphisms (and symplectomorphisms) of nilpotent Lie algebras as Lie algebra automorphisms $f: \mathfrak{g} \to \mathfrak{g}$, so that $\mathfrak{g}$ decomposes as a direct sum $\mathfrak{g} = \mathfrak{v}^s \oplus \mathfrak{v}^u$ of invariant stable and unstable subspaces.

Since the exponential map gives a diffeomorphism between the nilpotent Lie algebra $\mathfrak{g}$ and the associated simply connected nilpotent Lie group G, it follows that f induces a diffeomorphism $F: G \to G$, which is an Anosov diffeomorphism (or symplectomorphism). However, one has to check in each case for a given lattice $\Gamma \subset G$ whether F preserves Γ, so that F induces an Anosov diffeomorphism on the nilmanifold $M = \Gamma \backslash G$.

We want to give an explicit example of an Anosov symplectomorphism on a non-abelian nilpotent Lie algebra. According to a theorem of Malfait [Malf-00], a four-dimensional nilpotent Lie algebra admits an Anosov automorphism only if it is abelian. For $\mathfrak{nil}_4$, assuming that the Anosov automorphism is a symplectomorphism, we know this from other arguments, since we proved in Proposition 9.40 that $\mathfrak{nil}_4$ does not admit a Künneth structure. *A fortiori* it cannot admit an Anosov symplectomorphism. The only other non-abelian nilpotent Lie algebra of dimension 4 is $\mathfrak{nil}_3 \oplus \mathbb{R}$, which has lots of Künneth structures; compare Section 9.5 above. Even if this does not trivially contradict the existence of an Anosov symplectomorphism on $\mathfrak{nil}_3 \oplus \mathbb{R}$ – just think of the abelian case – it does make it unlikely, since an Anosov automorphism would induce a unique Künneth structure, which should not arise in an S^1-family. In any case, to find an interesting new example one has to consider non-abelian nilpotent Lie algebras in dimensions ≥ 6.

For the following construction we modify Example 2 on page 762 in Smale's paper [Sma-67]. We consider the six-dimensional nilpotent Lie group $G = Nil^3 \times Nil^3$ with Lie algebra $\mathfrak{g} = \mathfrak{nil}_3 \oplus \mathfrak{nil}_3$. As in Example 9.9 we think of $\mathfrak{nil}_3$ as $\mathfrak{n}_3$, consisting of the strictly upper-triangular matrices of the form

$$A = \begin{pmatrix} 0 & x & z \\ 0 & 0 & y \\ 0 & 0 & 0 \end{pmatrix} \quad \text{with} \quad x, y, z \in \mathbb{R} .$$

The corresponding basis vectors

$$X = \begin{pmatrix} 0 & 1 & 0 \\ 0 & 0 & 0 \\ 0 & 0 & 0 \end{pmatrix}, \quad Y = \begin{pmatrix} 0 & 0 & 0 \\ 0 & 0 & 1 \\ 0 & 0 & 0 \end{pmatrix}, \quad Z = \begin{pmatrix} 0 & 0 & 1 \\ 0 & 0 & 0 \\ 0 & 0 & 0 \end{pmatrix}$$

satisfy $[X, Y] = Z$. For the Lie algebra $\mathfrak{g}$ we have the direct sum of two copies of $\mathfrak{nil}_3$ with corresponding bases X_1, Y_1, Z_1 and X_2, Y_2, Z_2. Let $\alpha_1, \beta_1, \gamma_1$ and $\alpha_2, \beta_2, \gamma_2$ be the dual bases of one-forms. Then α_i, β_i are closed and

$$d\gamma_i = \beta_i \wedge \alpha_i , \quad i = 1, 2 .$$

For an arbitrary real number $\lambda > 1$ consider the linear isomorphism

$$f : \mathfrak{g} \longrightarrow \mathfrak{g}$$

defined in terms of the above basis by

$$\begin{aligned} X_1 &\longmapsto \lambda X_1 , & X_2 &\longmapsto \lambda^{-1} X_2 , \\ Y_1 &\longmapsto \lambda^{-2} Y_1 , & Y_2 &\longmapsto \lambda^2 Y_2 , \\ Z_1 &\longmapsto \lambda^{-1} Z_1 , & Z_2 &\longmapsto \lambda Z_2 . \end{aligned}$$

The map f is a linear Anosov isomorphism with stable subspace V^s spanned by $\{Y_1, Z_1, X_2\}$ and unstable subspace V^u spanned by $\{X_1, Y_2, Z_2\}$. Both V^s and V^u are in fact abelian Lie subalgebras of $\mathfrak{g}$.

We have

$$[f(X_i), f(Y_i)] = f(Z_i) = f([X_i, Y_i]) , \quad i = 1, 2 ,$$

hence f is a Lie algebra automorphism of $\mathfrak{g}$. Furthermore, the two-form

$$\omega = \alpha_1 \wedge \gamma_1 + \beta_1 \wedge \beta_2 + \alpha_2 \wedge \gamma_2$$

is closed and non-degenerate and satisfies $f^* \omega = \omega$. We have shown the following.

Proposition 9.43 *The linear automorphism $f : \mathfrak{g} \to \mathfrak{g}$ on $\mathfrak{g} = \mathfrak{nil}_3 \oplus \mathfrak{nil}_3$ is an Anosov symplectomorphism with respect to the symplectic structure ω.*

Since the Lie group G is simply connected and the exponential map $\exp\colon \mathfrak{g} \to G$ is a diffeomorphism by Theorem 9.14, it follows that f induces a unique Anosov diffeomorphism $F\colon G \to G$, which preserves the left-invariant symplectic form on G defined by ω.

We want to prove that F induces an Anosov symplectomorphism on a nil-manifold $M = \Gamma\backslash G$ associated to the Lie group $G = Nil^3 \times Nil^3$. For this we have to find a lattice $\Gamma \subset G$ which is preserved by F.

Consider the ring

$$R = \mathbb{Z}\left[\sqrt{3}\right] = \mathbb{Z} \oplus \sqrt{3}\mathbb{Z} \subset \mathbb{R}$$

with conjugation

$$\sigma\colon R \longrightarrow R,$$
$$a + \sqrt{3}b \longmapsto a - \sqrt{3}b \quad (a, b \in \mathbb{Z}).$$

The map σ is a ring automorphism. From now on we fix λ in the definition of the linear Anosov symplectomorphism f by

$$\lambda = 2 + \sqrt{3} \in R.$$

Then

$$\lambda^{-1} = 2 - \sqrt{3} \in R,$$

i.e. $\lambda^{-1} = \sigma(\lambda)$.

Consider the set $\mathfrak{nil}_3(R)$ of elements of $\mathfrak{nil}_3$ of the form

$$A = \begin{pmatrix} 0 & x & z \\ 0 & 0 & y \\ 0 & 0 & 0 \end{pmatrix}, \quad x, y, z \in R.$$

Since R is dense in $\mathbb{R}$, the set $\mathfrak{nil}_3(R)$ is dense in $\mathfrak{nil}_3$. However, the subset

$$(\mathfrak{nil}_3 \oplus \mathfrak{nil}_3)(R)$$

of $\mathfrak{nil}_3 \oplus \mathfrak{nil}_3$ of elements of the form $(A, \sigma(A))$, with $A \in \mathfrak{nil}_3(R)$ and

$$\sigma(A) = \begin{pmatrix} 0 & \sigma(x) & \sigma(z) \\ 0 & 0 & \sigma(y) \\ 0 & 0 & 0 \end{pmatrix},$$

is a discrete lattice in the Lie algebra $\mathfrak{g} = \mathfrak{nil}_3 \oplus \mathfrak{nil}_3$, closed under commutators. It follows that the image of $(\mathfrak{nil}_3 \oplus \mathfrak{nil}_3)(R)$ under the exponential map is a lattice Γ in the Lie group $G = Nil^3 \times Nil^3$.

Note that the linear isomorphism $f\colon \mathfrak{g} \to \mathfrak{g}$ maps $(\mathfrak{nil}_3 \oplus \mathfrak{nil}_3)(R)$ onto itself, because R is a ring and $\lambda, \lambda^{-1} \in R$. As a consequence, the Anosov symplectomorphism $F\colon G \to G$ maps the lattice Γ onto itself, and we get the following.

Proposition 9.44 *The linear Anosov symplectomorphism* $f: \mathfrak{g} \to \mathfrak{g}$ *induces an Anosov symplectomorphism on the six-dimensional symplectic nilmanifold* $M = \Gamma \backslash G$.

According to Theorem 9.25, the nilmanifold M is a principal T^2-bundle over T^4.

9.7 Solvmanifolds

Going beyond nilmanifolds, one can try to generalise some of the discussion in this chapter to the larger class of solvmanifolds, hoping to obtain more examples of Künneth structures in this way.

Definition 9.45 Let G be a simply connected solvable Lie group. A *lattice* $\Gamma \subset G$ is a discrete, co-compact subgroup. The quotient manifold $M = \Gamma \backslash G$ is called a *solvmanifold*.

As before, we can define the notions of symplectic structures, of Lagrangian foliations and of Künneth structures on solvable Lie algebras and on solvmanifolds. We highlight some of the differences between solvmanifolds and nilmanifolds.

(i) A simply connected solvable Lie group is still diffeomorphic to $\mathbb{R}^n$, but the exponential map in general is not a diffeomorphism, hence the analogue of Theorem 9.14 does not hold for general solvable Lie groups. See [Kna-96, Chapter I, Corollary 1.103].

(ii) There is no theorem comparable to Malcev's Theorem (Theorem 9.16) that guarantees the existence of lattices in simply connected solvable Lie groups.

(iii) The statement of Theorem 9.35 due to Baues and Cortés concerning the existence of Lagrangian foliations for every symplectic nilpotent Lie algebra does not hold for general solvable Lie algebras. A counterexample can be found in Proposition 3.21 in [BC-16].

Notwithstanding these difficulties, in some simple cases one can still find explicit Künneth structures on suitably chosen solvable Lie algebras.

Example 9.46 We want to consider a solvable Lie algebra whose associated simply connected solvable Lie group is known to admit lattices. Let $\mathfrak{g} = \mathfrak{sol}_3 \oplus \mathbb{R}$

be the Lie algebra with basis e_1, e_2, e_3, e_4 and non-zero commutators

$$[e_1, e_3] = -e_1 \,,$$
$$[e_2, e_3] = e_2 \,.$$

This Lie algebra is indeed solvable, for we have

$$[\mathfrak{g}, \mathfrak{g}] = \mathrm{span}\{e_1, e_2\}$$

and

$$[[\mathfrak{g}, \mathfrak{g}], [\mathfrak{g}, \mathfrak{g}]] = 0 \,.$$

However, the Lie algebra is not nilpotent, because

$$\mathfrak{g}_{(k)} = [\mathfrak{g}, \mathfrak{g}] \neq 0 \quad \forall k \geq 2 \,.$$

For the dual basis of 1-forms we have

$$d\alpha_1 = \alpha_1 \wedge \alpha_3 \,,$$
$$d\alpha_2 = \alpha_3 \wedge \alpha_2 \,,$$
$$d\alpha_3 = d\alpha_4 = 0 \,.$$

We make the following general ansatz for a symplectic form

$$\omega = \sum_{i<j} A_{ij} \alpha_i \wedge \alpha_j \,.$$

Then

$$0 = d\omega = A_{14} \alpha_1 \wedge \alpha_3 \wedge \alpha_4 + A_{24} \alpha_3 \wedge \alpha_2 \wedge \alpha_4$$

implies $A_{14} = A_{24} = 0$, hence the symplectic form is

$$\omega = A_{12} \alpha_1 \wedge \alpha_2 + A_{13} \alpha_1 \wedge \alpha_3 + A_{23} \alpha_2 \wedge \alpha_3 + A_{34} \alpha_3 \wedge \alpha_4 \,.$$

The condition $\omega^2 \neq 0$ implies $A_{12} A_{34} \neq 0$. Conversely, whenever this is satisfied, we have a symplectic form. Consider

$$\mathfrak{f} = \mathrm{span}\{e_1, e_3 - (A_{13}/A_{12}) e_2\} \,,$$
$$\mathfrak{h} = \mathrm{span}\{e_2, e_4\} \,.$$

Then $\mathfrak{f}$ and $\mathfrak{h}$ are a pair of complementary Lagrangian foliations for the above symplectic form on $\mathfrak{sol}_3 \oplus \mathbb{R}$. This yields Künneth structures on all solvmanifolds associated to the solvable Lie group $Sol^3 \times \mathbb{R}$.

Notes for Chapter 9

1. The material of Section 9.4 is largely taken from [Ham-04, Ham-19].

2. The paper [FPPS-04] contains other examples in the style of the eight-dimensional example discussed in Section 9.5. The paper by Andrada–Dotti [AD-06] contains constructions of hypersymplectic structures on some three-step nilpotent Lie algebras.

3. Andrada [And-06] gives a characterisation of hypersymplectic Lie algebras.

4. A modification of Smale's Example 2 in [Sma-67] that is very similar to our discussion in Section 9.6 was discussed by Shub [Shub-69]. Whereas our interest is primarily in having an invariant symplectic structure for an Anosov diffeomeorphism, so that we obtain an induced Künneth structure, Shub's priority was to construct an example of an Anosov diffeomorphism, not necessarily symplectic, that descends to a genuine infra-nilmanifold that is not a nilmanifold.

10

Four-manifolds

In this chapter we discuss Künneth geometry in real dimension 4. Since in dimension 2 Künneth geometry is essentially Lorentz geometry, dimension 4 is really the first interesting case. For at least two reasons, it is also a very special case. First, it is possible to classify almost Künneth structures in terms of classical invariants. Second, four-dimensional symplectic geometry is very subtle, and symplectic structures in this dimension are constrained by their relation with Seiberg–Witten gauge theory. We shall see that this makes it likely that Künneth four-manifolds may be classified, although we do not achieve that goal here, except in the hypersymplectic case.

Throughout this chapter we will use not only the material developed in earlier chapters of this book, but also the tools of modern four-dimensional geometry and topology. In particular we will use results from gauge theory. A good reference for both the basics of four-dimensional differential topology and results from Donaldson theory is the book by Donaldson and Kronheimer [DK-90]. In fact, very little Donaldson theory will be used in this chapter. We will make more use of results from Seiberg–Witten theory, for which we refer to the book by Morgan [Mor-96] and the second author's Bourbaki lecture [Kot-97a] on Taubes's work.

10.1 Classical Invariants

Since we are interested in Künneth structures on compact manifolds, we recall here some standard facts about the algebraic topology of closed smooth four-manifolds. We may and do restrict to the oriented category because an almost Künneth structure defines an orientation.

The most basic homotopy invariant of a connected manifold M is of course its fundamental group $\pi_1(M)$. While the fundamental groups of manifolds of dimensions ≤ 3 are very special, in dimension 4 there is no longer any constraint. The following result goes back to Dehn [Deh-12].

Theorem 10.1 *Every finitely presentable group is the fundamental group of a closed oriented smooth four-manifold.*

One way to prove this is to take a connected sum of copies of $S^1 \times S^3$, one for each generator of a finite presentation of the group in question, and then perform elementary surgery on a finite number of embedded circles representing the finitely many relations in the presentation.

Next, we have the integral homology and cohomology groups, $H_i(M; \mathbb{Z})$ and $H^i(M; \mathbb{Z})$, respectively. These groups determine the homology and cohomology groups with other coefficients via the universal coefficient theorem, and they satisfy Poincaré duality.

For a connected closed oriented four-manifold, $H_0(M; \mathbb{Z})$ and $H_4(M; \mathbb{Z})$ are both infinite cyclic, with the latter generated by the fundamental class $[M]$ of an orientation. The first homology $H_1(M; \mathbb{Z})$ is just the Abelianisation of the fundamental group $\pi_1(M)$. The second homology $H_2(M; \mathbb{Z})$ is much more interesting. It is not determined by the fundamental group alone and, moreover, carries a bilinear form called the intersection form of M. This can be defined geometrically by intersecting pairs of transversal submanifolds. However, it is technically more convenient to pass to cohomology and use the cup product. Then the intersection form becomes

$$Q_M : H^2(M; \mathbb{Z}) \times H^2(M; \mathbb{Z}) \longrightarrow \mathbb{Z}$$
$$([\phi], [\psi]) \longmapsto \langle \phi \cup \psi, [M] \rangle .$$

By Poincaré duality this symmetric bilinear form is non-degenerate if we mod out the torsion in $H^2(M; \mathbb{Z})$, or if we take coefficients in $\mathbb{R}$, say. Over $\mathbb{R}$ a symmetric bilinear form is determined by its rank and its signature. Here the rank is just the second Betti number $b_2(M)$, and if we write $b_2^\pm(M)$ for the maximal dimensions of a positive, respectively negative, definite subspace with respect to Q_M in $H^2(M; \mathbb{R})$, then we have

$$b_2(M) = b_2^+(M) + b_2^-(M) ,$$

and the signature is given by

$$\sigma(M) = b_2^+(M) - b_2^-(M) .$$

By the Hirzebruch signature formula, the signature is given by evaluating a combination of Pontryagin classes, in this case just the first Pontryagin class:

$$\sigma(M) = \frac{1}{3} \langle p_1(TM), [M] \rangle . \tag{10.1}$$

Now, over $\mathbb{Z}$, symmetric bilinear forms are not classified by their rank and signature. At the very least, one has to treat separately the so-called even forms, for which $Q(x, x)$ is even for all x, and the odd forms, which by definition are all the ones that are not even. The parity of the intersection form of a smooth manifold M is detected by its second Stiefel–Whitney class $w_2(TM)$, because this is a characteristic element for the intersection form, i.e.

$$Q_M(x, x) \equiv Q_M(w_2(TM), x) \mod 2 \quad \forall x \in H^2(M; \mathbb{Z}) .$$

For spin manifolds $w_2(TM) = 0$ by definition, and so the intersection form is even. The converse is true if $H^2(M; \mathbb{Z})$ is free of two-torsion.

Although there is no classification of definite symmetric bilinear forms over $\mathbb{Z}$, in the indefinite case one has the Hasse–Minkowski classification, giving a complete list of all indefinite symmetric bilinear forms over $\mathbb{Z}$, up to isomorphism over $\mathbb{Z}$. A proof can be found, for example, in the booklet by Serre [Ser-79].

Theorem 10.2 (Hasse–Minkowski classification) *If Q_M is indefinite, then it is equivalent either to $p(1) \oplus q(-1)$ with $p, q > 0$ if it is odd, or to $aH \oplus bE_8$ with $a \geq 1$ and $b \in \mathbb{Z}$ if it is even.*

Here H is the intersection form of $S^2 \times S^2$ given by

$$H = \begin{pmatrix} 0 & 1 \\ 1 & 0 \end{pmatrix}$$

and E_8 is the incidence matrix of the Dynkin diagram of the exceptional Lie group E_8:

$$E_8 = \begin{pmatrix} -2 & 1 & 0 & 0 & 0 & 0 & 0 & 0 \\ 1 & -2 & 1 & 0 & 0 & 0 & 0 & 0 \\ 0 & 1 & -2 & 1 & 0 & 0 & 0 & 0 \\ 0 & 0 & 1 & -2 & 1 & 0 & 0 & 0 \\ 0 & 0 & 0 & 1 & -2 & 1 & 0 & 1 \\ 0 & 0 & 0 & 0 & 1 & -2 & 1 & 0 \\ 0 & 0 & 0 & 0 & 0 & 1 & -2 & 0 \\ 0 & 0 & 0 & 0 & 1 & 0 & 0 & -2 \end{pmatrix} .$$

Note that, since $\sigma(H) = 0$ and $\sigma(E_8) = -8$, the Hasse–Minkowski classification implies the following.

Lemma 10.3 *If Q_M is even then $\sigma(M) \equiv 0 \pmod 8$.*

This statement is much more elementary than the actual classification and

can be proved more easily. We have deduced it from the classification just for
convenience.

Suppose now that our smooth manifold M has an almost complex structure
J. Then $c_1(M, J)$ is a lift of $w_2(TM)$ to an integral class. Moreover, the relation
between Chern and Pontryagin classes gives

$$p_1(TM) = c_1^2(M, J) - 2c_2(M, J) .$$

Since the top Chern class $c_2(M, J)$ is the Euler class of TM, we find

$$c_1^2(M, J) = 2e(TM) + p_1(TM) .$$

Evaluating on the fundamental class of M and taking into account the signature
formula (10.1), we find the so-called Wu formula:

$$c_1^2(M, J) = 2\chi(M) + 3\sigma(M) , \qquad (10.2)$$

where on the left-hand side we suppressed the evaluation on $[M]$ in the notation.

An important candidate for the existence of Künneth structures is the so-
called $K3$ manifold, which is the oriented smooth manifold underlying a complex $K3$ surface. Here is a way of defining such a $K3$ surface.

Example 10.4   Consider a quartic smooth surface $S \subset \mathbb{C}P^3$, i.e. the zero locus
of a generic polynomial of degree 4 in the homogeneous coordinates. This is
a projective-algebraic, hence Kähler, complex surface, whose first Chern class
vanishes by the adjunction formula. It follows that it is spin, and one can easily
calculate $\chi(S) = 24$ and $\sigma(S) = -16$. By the Hasse–Minkowski classification
the intersection form of S is $Q_S = 3H \oplus 2E_8$. Note also that S is simply
connected by the Lefschetz Hyperplane Theorem.

10.2 Almost Künneth Structures

We begin by characterising the existence or non-existence of almost Künneth
structures in terms of classical invariants. We state the result only for orientable
structures. Since orientability can always be achieved by passage to a suitable
double covering, this is not a serious restriction.

Theorem 10.5   *On a closed oriented smooth four-manifold M the following
conditions are equivalent:*

  (i) *M admits an orientable almost Künneth structure,*
 (ii) *there is an oriented real rank 2 bundle $L \to M$ with $TM \cong L \oplus \overline{L}$,*

(iii) *M is spin with* $2\chi(M) + 3\sigma(M) = 0$,

(iv) *M admits an almost complex structure J with* $c_1(M, J) = 0$.

Proof If M has an orientable almost Künneth structure, then recalling Lemma 2.26 and $\epsilon(2) = -1$, we have $TM \cong L \oplus \overline{L}$, with L an oriented real vector bundle of rank 2. Conversely, assume that TM admits a splitting as $L \oplus \overline{L}$, with L oriented. Then M also admits an almost complex structure J with $c_1(M, J) = 0$ for which L and $\overline{L}$ are totally real. This proves the equivalence of the first two conditions.

We now show the equivalence of the first two conditions with the third one. First, if M admits an orientable almost Künneth structure, then its first Chern class vanishes by Theorem 2.38; cf. (2.2). This shows that M is spin and that $2\chi(M) + 3\sigma(M) = c_1^2(M) = 0$ by the Wu formula (10.2). Conversely, assume that M is spin with $2\chi(M) + 3\sigma(M) = 0$. The spin assumption implies that the intersection form Q_M is even.

If Q_M is indefinite, then by the Hasse–Minkowski classification recalled in Theorem 10.2, it splits off a hyperbolic summand H. In this summand, and *a fortiori* in the whole intersection form, all even integers are realised as self-intersection numbers of integral classes. Thus we can certainly find a $c \in H^2(M; \mathbb{Z})$ such that $c^2 = -\chi(M) = -2 + 2b_1(M) - b_2(M)$ (where the right-hand side is even because the intersection form is). Let $L \to M$ be the real oriented rank 2 bundle with Euler class c. Then $L \oplus \overline{L}$ is oriented with vanishing second Stiefel–Whitney class and with the same Euler and Pontryagin classes as TM. It follows that $TM \cong L \oplus \overline{L}$, since M has the homotopy type of a CW-complex and over a four-complex real rank 4 vector bundles are determined by their characteristic classes. (This result goes back to Pontryagin; cf. the corollary on p. 674 of [DW-59].)

If the intersection form Q_M is definite, then Donaldson's Theorem about smooth manifolds with definite intersection forms (see [Don-86]) shows that the second Betti number vanishes. Therefore the signature vanishes, and the assumptions then imply that the Euler characteristic vanishes as well. As M is assumed to be spin, it follows that it is parallelisable, again by the corollary on p. 674 of [DW-59], and therefore clearly almost Künneth. This completes the proof of the equivalence of the first three conditions.

We have already noted above that the first condition implies the fourth one, and that the fourth one implies the third. Thus all four conditions are indeed equivalent. □

Remark 10.6 It is not possible to avoid the use of Donaldson's Theorem in the above proof. Without invoking this theorem, one does not know that there are no manifolds with non-trivial positive-definite intersection form, for which the above equivalence would break down. The same issue makes it necessary to invoke Donaldson's Theorem when translating Wu's criterion for the existence of an almost complex structure into a condition on the Betti numbers only.

As a very special case of this theorem we see that the $K3$ manifold admits an almost Künneth structure, although its Euler characteristic does not vanish. Further examples can be generated by considering connected sums of several copies of the $K3$ manifold, of $S^1 \times S^3$, T^4, etc. One can also show that the existence of an orientable almost Künneth structure on a four-manifold does not constrain the fundamental group.

Corollary 10.7 *Every finitely presentable group is the fundamental group of a closed smooth four-manifold with an orientable almost Künneth structure.*

Proof Let N be any smooth closed oriented spin four-manifold with the given group as its fundamental group. The sketch given above for the proof of Dehn's Theorem (Theorem 10.1) provides such an N since one can carry it out within the spin category.

Next let M be the following connected sum:

$$M = N \# k(S^2 \times S^2) \# \ell K3$$

for non-negative k and ℓ. Then M and N are both spin, with the same fundamental group, and satisfy

$$\chi(M) = \chi(N) + 2k + 22\ell$$

and

$$\sigma(M) = \sigma(N) - 16\ell \, .$$

Therefore

$$2\chi(M) + 3\sigma(M) = 2\chi(N) + 3\sigma(N) + 4k - 4l \, .$$

Note that $\sigma(N)$ is divisible by 8 by Lemma 10.3, and $\chi(N)$ is even. So $2\chi(N) + 3\sigma(N)$ is certainly divisible by 4, so that choosing k and ℓ suitably we can arrange $2\chi(M) + 3\sigma(M) = 0$. Now Theorem 10.5 shows that this implies that M has an orientable almost Künneth structure. □

10.3 The Integrable Case

Imposing the integrability of an almost Künneth structure results in very serious additional constraints. We begin to explore these constraints using the work of Taubes on the Seiberg–Witten invariants of symplectic four-manifolds.

Proposition 10.8 *Let M be a closed oriented four-manifold M with a Künneth structure.*

(i) *If $b_2^+(M) > 1$, or the Künneth structure is orientable, then the first Chern class vanishes.*

(ii) *If $b_2^+(M) > 1$, then every symplectic form on M, even if it does not have a Künneth structure, must have trivial first Chern class.*

Proof If the Künneth structure is orientable, then by Theorem 2.38 its first Chern class vanishes. Without the orientability assumption, one concludes that the first Chern class is a two-torsion class. If $b_2^+(M) > 1$, then results of Taubes [Tau-95] about the Seiberg–Witten invariants of symplectic four-manifolds imply, first, that this two-torsion class vanishes, and, second, that this vanishing of the first Chern class is also true for every other symplectic structure on M; cf. [Kot-97a, Corollary 4.3]. □

Corollary 10.9 *There exist closed oriented four-manifolds which admit symplectic forms and almost Künneth structures, but do not admit any Künneth structure.*

Proof The manifolds $E(2n)$, the standard simply connected spin elliptic surfaces without multiple fibres over $\mathbb{C}P^1$ satisfy the criterion of Theorem 10.5, and therefore admit orientable almost Künneth structures. They also have obvious symplectic, in fact Kähler, structures. However, for all $n > 1$ the Kähler structures do not have trivial first Chern class, and so Proposition 10.8 rules out the existence of Künneth structures. □

Proposition 10.8 shows that in order to investigate Künneth structures on smooth four-manifolds one needs to come to grips with the class of symplectic four-manifolds which have vanishing, or at least torsion, first Chern class. In the literature such manifolds are often called symplectic Calabi–Yau, or SCY, manifolds, because Kähler manifolds with trivial first Chern classes admit Calabi–Yau metrics. Examples of SCY four-manifolds are very scarce, and all known examples are finitely covered either by the $K3$ manifold or by a T^2-bundle over T^2. We will discuss T^2-bundles over T^2 in detail in Section 10.6 below, where we prove that many of them admit Künneth structures.

We have seen in Theorem 10.5 that the $K3$ manifold does have an orientable almost Künneth structure, but we have so far been unable to decide whether an integrable one can exist or not. We do, however, know the following weaker statement.

Proposition 10.10 *The K3 manifold does not have a flat Künneth structure. In particular, it does not admit a Kähler–Künneth structure.*

Proof A flat Künneth connection on M gives the tangent bundle TM the structure of a flat vector bundle, so that its real Pontryagin classes must vanish. But the first Pontryagin class of the $K3$ manifold is non-trivial since its signature does not vanish; cf. (10.1). The second claim follows from Theorem 7.27, asserting that any Kähler–Künneth structure is flat.   □

This proposition does not quite rule out the existence of a Künneth structure for which the symplectic form is the Kähler form of a complex $K3$ surface, because one could still have such a structure for which the Lagrangian foliations are not parallel, so that it is not Kähler–Künneth.

We can rule out the existence of hypersymplectic structures on the $K3$ manifold, and, in the process, classify all closed hypersymplectic four-manifolds.

Proposition 10.11 *A closed oriented four-manifold admits a hypersymplectic structure if and only if it is T^4 or a nilmanifold for $Nil^3 \times \mathbb{R}$.*

Proof By our discussion in Subsection 8.1.2 of Chapter 8, a closed oriented four-manifold with a hypersymplectic structure is holomorphic symplectic. Therefore, according to a result of Kodaira, it is T^4, a primary Kodaira surface, or a $K3$ surface; see [BPV-84]. Clearly T^4 carries the standard hypersymplectic structure of $\mathbb{R}^4$.

By a result of Wall [Wall-86], the primary Kodaira surfaces are precisely the nilmanifolds of $Nil^3 \times \mathbb{R}$. In Section 9.5 of Chapter 9 we showed that $Nil^3 \times \mathbb{R}$ carries a left-invariant hypersymplectic structure, which therefore descends to all the nilmanifolds for this group.

By our discussion in Subsection 8.1.1 of Chapter 8, a hypersymplectic structure also defines two symplectic pairs, for example $\omega \pm \alpha$. However, in dimension 4, the two symplectic forms making up a symplectic pair induce opposite orientations, and therefore a four-manifold with such a structure is symplectic for both choices of orientation. But the $K3$ manifold endowed with the non-complex orientation cannot be symplectic, because it has vanishing Seiberg–Witten invariants. This follows from the existence of smoothly embedded spheres whose self-intersection number is positive for the non-complex orientation; cf. [Kot-97b].   □

10.4 Constraints on Symplectic Calabi–Yau Four-manifolds

We now discuss in more detail the topology of the class of symplectic four-manifolds that are candidates for supporting orientable Künneth structures.

Definition 10.12 A closed oriented smooth four-manifold is called *symplectic Calabi–Yau*, or *SCY*, if it admits an orientation-compatible symplectic form such that the first Chern class of the associated almost complex structure vanishes.

These manifolds are of course spin, since the first Chern class reduces mod 2 to the second Stiefel–Whitney class. The deepest and most important result about this class of manifolds is the following theorem of Bauer [Bau-08] and, independently, Li [Li-06]. The proof consists of a subtle argument with refined Seiberg–Witten theory on spin manifolds, contrasted with Taubes's results on Seiberg–Witten invariants of symplectic manifolds.

Theorem 10.13 (Bauer [Bau-08], Li [Li-06]) *Let M be a closed symplectic Calabi–Yau four-manifold. Then the Betti numbers satisfy one of the following conditions:*

(i) *either $b_1(M) = 0$, $\chi(M) = 24$ and $\sigma(M) = -16$,*
(ii) *or $2 \leq b_1(M) \leq 4$ and $\chi(M) = \sigma(M) = 0$.*

The first case is realised by the $K3$ manifold, and all possibilities in the second case are realised by T^2-bundles over T^2. It is natural to conjecture that, up to finite coverings, these are in fact the only manifolds admitting SCY structures, and therefore the only candidates for smooth four-manifolds that may possibly admit integrable orientable Künneth structures.

The theorem shows that the fundamental groups of SCY four-manifolds are severely restricted, in sharp contrast with the case of almost Künneth four-manifolds treated in Corollary 10.7. Beyond the obvious restrictions imposed by the possible values of the first Betti number, there are also some slightly more subtle restrictions. In the case of vanishing first Betti number we have the following.

Corollary 10.14 *Let M be a closed symplectic Calabi–Yau four-manifold with $b_1(M) = 0$. Then $\pi_1(M)$ has no proper subgroups of finite index.*

Proof Suppose the conclusion were false. Then lifting the SCY structure to the corresponding finite covering manifold would produce a counterexample to the theorem because signature and Euler characteristic are multiplicative in finite coverings. □

The only example that is known for this result is in fact simply connected. With the stronger assumption of simple connectivity, we have the following.

Corollary 10.15 *Let M be a closed symplectic Calabi–Yau four-manifold* with $\pi_1(M) = 0$. *Then M is orientation-preservingly homeomorphic to the K3 manifold.*

Proof The first part of Theorem 10.13 shows that M has the same rational or real intersection form as the $K3$ manifold. However, M is spin, and so its intersection form is even. By the Hasse–Minkowski classification, Theorem 10.2, Q_M is – over $\mathbb{Z}$ – equivalent to the intersection form of the $K3$ manifold. Thus M and $K3$ are orientation-preservingly homeomorphic by Freedman's classification of simply connected four-manifolds. □

For the case of positive first Betti number an immediate consequence of Theorem 10.13 is the following.

Corollary 10.16 *Every closed SCY four-manifold with positive first Betti number is parallelisable.*

Proof The vanishing of the Stiefel–Whitney classes and of the Euler characteristic and signature together imply triviality of the tangent bundle by a result going back to Pontryagin; cf. the corollary on p. 674 of [DW-59]. □

Parallelisability imposes various restrictions on the fundamental group, some of which were explored in the work of Johnson–Kotschick [JK-93] and of Johnson–Walton [JW-00]. Here we will make explicit only one specific statement that can be phrased in terms of the invariants p and q surveyed in [Kot-94].

We first recall the definitions. For a finitely presentable group Γ one can consider the collection of all connected closed oriented four-manifolds M with fundamental group Γ. This collection is non-empty by Dehn's Theorem (Theorem 10.1). The so-called Hausmann–Weinberger invariant $q(\Gamma)$ is the infimum over the Euler characteristics of all these manifolds. Similarly, $p(\Gamma)$ is the infimum over all the differences

$$\chi(M) - |\sigma(M)|$$

for M as above with $\pi_1(M) = \Gamma$. By their very definitions, these invariants satisfy the following straightforward inequalities; cf. [Kot-94, Theorem 2.5]:

$$2 - 2b_1(\Gamma) \le p(\Gamma) \le q(\Gamma) . \tag{10.3}$$

It is known that all pairs of integers (x, y) with $x \le y$ are realised as $(p(\Gamma), q(\Gamma))$ by infinitely many pairwise non-isomorphic finitely presentable groups Γ; cf. [Kot-94, Theorem 6.1].

For the fundamental groups of SCY manifolds we now prove the following.

Proposition 10.17 *If Γ is the fundamental group of a closed SCY four-manifold with positive first Betti number, then*

$$p(\Gamma) = q(\Gamma) = 0 .$$

Proof On the one hand, by Theorem 10.13 the SCY manifold M with $\pi_1(M) = \Gamma$ has vanishing Euler characteristic. Therefore, the second inequality in (10.3) gives

$$p(\Gamma) \le q(\Gamma) \le \chi(M) = 0 . \tag{10.4}$$

On the other hand, $b_1(\Gamma) > 0$ implies that Γ has subgroups $\Gamma' \subset \Gamma$ of unbounded finite index k. If $p(\Gamma)$ were negative, then there would be a closed smooth four-manifold N with fundamental group Γ and with

$$\chi(N) - |\sigma(N)| < 0 .$$

Passing to the covering N' with fundamental group Γ', the multiplicativity of the signature and the Euler characteristic in coverings would give

$$p(\Gamma') \le \chi(N') - |\sigma(N')| = k(\chi(N) - |\sigma(N)|) < 0 .$$

Since k is unbounded, $p(\Gamma')$ could be made as negative as one wants to, by choosing Γ' appropriately. However, lifting the SCY structure from N to N', Theorem 10.13 implies $b_1(\Gamma') \le 4$ for all these subgroups Γ'. Therefore the first inequality in (10.3) gives

$$p(\Gamma') \ge 2 - 2b_1(\Gamma') \ge -6 .$$

This contradiction proves $p(\Gamma) \ge 0$. Together with (10.4), this finishes the proof. □

10.5 Künneth–Einstein Structures

In this section we extend Theorem 7.19 to dimension 4, by showing that every closed Künneth–Einstein four-manifold is Ricci-flat. The proof involves a discussion of self-duality of two-forms with respect to the Künneth metric of signature $(2, 2)$. Up to certain sign changes this parallels the standard discussion of self-duality in four-dimensional Riemannian geometry.

Let $(M, \omega, \mathcal{F}, \mathcal{G})$ be a Künneth four-manifold and g the induced metric of neutral signature defined by

$$g(X, Y) = \omega(IX, Y) .$$

Lemma 10.18 *We have $g(\omega, \omega) = -2$.*

Proof For a point $p \in M$ let f_1, f_2, g_1, g_2 be a symplectic basis of $T_p M$ adapted to the foliations so that

$$\omega(f_1, g_1) = 1 = \omega(f_2, g_2) \, .$$

Then also

$$g(f_1, g_1) = 1 = g(f_2, g_2)$$

and all other scalar products between the basis vectors vanish. Let $\alpha_1, \alpha_2, \beta_1, \beta_2$ be the dual basis of $T_p^* M$. Since

$$\omega = \alpha_1 \wedge \beta_1 + \alpha_2 \wedge \beta_2 \, ,$$

it follows that

$$\begin{aligned} g(\omega, \omega) &= g(f_1 \wedge g_1, f_1 \wedge g_1) + g(f_2 \wedge g_2, f_2 \wedge g_2) \\ &= -g(f_1, g_1)^2 - g(f_2, g_2)^2 \\ &= -2 \, . \end{aligned}$$

$\square$

Remark 10.19   Although the metric g is uniquely defined on vectors or one-forms only up to an overall choice of a sign, it is unique on two-forms, since the evaluation on two-forms involves the product of two scalar products on one-forms.

As always, we consider the four-manifold M to be equipped with its symplectic orientation defined by the volume form $\omega \wedge \omega$. The orientation together with g yields a volume form dvol_g.

Lemma 10.20   *We have* $\omega \wedge \omega = 2 \mathrm{dvol}_g$.

Proof   For $i = 1, 2$ let

$$\gamma_i^{\pm} = \frac{1}{\sqrt{2}} (\alpha_i \pm \beta_i) \, .$$

Then $\gamma_1^+, \gamma_1^-, \gamma_2^+, \gamma_2^-$ are an orthonormal basis for the scalar product g_p on $T_p^* M$ with

$$g(\gamma_i^{\pm}, \gamma_i^{\pm}) = \pm 1 \, .$$

Furthermore,

$$\gamma_i^+ \wedge \gamma_i^- = -\alpha_i \wedge \beta_i \, ,$$

and hence

$$\gamma_1^+ \wedge \gamma_1^- \wedge \gamma_2^+ \wedge \gamma_2^- = \alpha_1 \wedge \beta_1 \wedge \alpha_2 \wedge \beta_2$$
$$= \frac{1}{2} \omega \wedge \omega .$$

In particular, $(\gamma_1^+, \gamma_1^-, \gamma_2^+, \gamma_2^-)$ is a positively oriented basis, and since it is orthonormal we conclude

$$\mathrm{dvol}_g = \gamma_1^+ \wedge \gamma_1^- \wedge \gamma_2^+ \wedge \gamma_2^-$$

and therefore $\omega \wedge \omega = 2\mathrm{dvol}_g$. $\square$

The metric g and the orientation define a Hodge star operator

$$*: \Lambda^2 T^* M \longrightarrow \Lambda^2 T^* M$$

by

$$\alpha \wedge *\beta = g(\alpha, \beta)\mathrm{dvol}_g \quad \forall \alpha, \beta \in \Lambda^2 T^* M .$$

Since $** = \mathrm{Id}$, we can decompose the two-forms on M into self-dual and anti-self-dual forms:

$$\Lambda^2 T^* M = \Lambda^+ \oplus \Lambda^- , \quad *|_{\Lambda^\pm} = \pm \mathrm{Id}|_{\Lambda^\pm} .$$

This decomposition can be interpreted in terms of the bigrading on forms induced from the splitting

$$TM = T\mathcal{F} \oplus T\mathcal{G} .$$

Lemma 10.21 *The vector bundle of anti-self-dual two-forms is given by*

$$\Lambda^- = \Lambda^{2,0} \oplus \mathbb{R}\omega \oplus \Lambda^{0,2} .$$

Proof For our choice of oriented orthonormal basis we have

$$\alpha_i = \frac{1}{\sqrt{2}}(\gamma_i^+ + \gamma_i^-) ,$$
$$\beta_i = \frac{1}{\sqrt{2}}(\gamma_i^+ - \gamma_i^-) .$$

It follows that

$$\omega = -\gamma_1^+ \wedge \gamma_1^- - \gamma_2^+ \wedge \gamma_2^- ,$$
$$\alpha_1 \wedge \alpha_2 = \frac{1}{2}(\gamma_1^+ \wedge \gamma_2^+ + \gamma_1^+ \wedge \gamma_2^- + \gamma_1^- \wedge \gamma_2^+ + \gamma_1^- \wedge \gamma_2^-) ,$$
$$\beta_1 \wedge \beta_2 = \frac{1}{2}(\gamma_1^+ \wedge \gamma_2^+ - \gamma_1^+ \wedge \gamma_2^- - \gamma_1^- \wedge \gamma_2^+ + \gamma_1^- \wedge \gamma_2^-) .$$

A calculation shows that

$$*(\gamma_1^+ \wedge \gamma_2^+) = -\gamma_1^- \wedge \gamma_2^-\,,$$
$$*(\gamma_1^+ \wedge \gamma_1^-) = -\gamma_2^+ \wedge \gamma_2^-\,,$$
$$*(\gamma_1^+ \wedge \gamma_2^-) = -\gamma_1^- \wedge \gamma_2^+\,.$$

Together with $** = \mathrm{Id}$ this implies that ω, $\alpha_1 \wedge \alpha_2$ and $\beta_1 \wedge \beta_2$ are indeed anti-self-dual. $\square$

Let R denote the Riemann curvature tensor of g and $\mathcal{R}$ the curvature operator

$$\mathcal{R}\colon \Lambda^2 TM \longrightarrow \Lambda^2 TM\,,$$

defined by

$$g(\mathcal{R}(x \wedge y), u \wedge v) = g(R(x,y)v, u)\,.$$

The metric g defines an isomorphism

$$TM \longrightarrow T^*M\,, \tag{10.5}$$

that maps

$$f_i \mapsto \beta_i\,, \quad g_i \mapsto \alpha_i\,.$$

With the induced isomorphism on two-forms we get a curvature operator

$$\mathcal{R}\colon \Lambda^2 T^*M \longrightarrow \Lambda^2 T^*M\,.$$

Lemma 10.22 *The curvature tensor $\mathcal{R}$ vanishes identically on $\Lambda^{2,0}$ and $\Lambda^{0,2}$.*

Proof This is true because the tangent bundle TM is flat along the leaves of $\mathcal{F}$ and $\mathcal{G}$. Hence

$$\mathcal{R}(f_1 \wedge f_2) = 0 = \mathcal{R}(g_1 \wedge g_2)$$

and

$$\mathcal{R}(\beta_1 \wedge \beta_2) = 0 = \mathcal{R}(\alpha_1 \wedge \alpha_2)\,.$$

$\square$

Lemma 10.23 *Suppose that g is Einstein with $\mathrm{Ric} = \lambda g$. Then*

$$\mathcal{R}(\omega) = \lambda \omega\,.$$

It follows that $\mathcal{R}$ restricts to a linear map on

$$\Lambda^- = \Lambda^{2,0} \oplus \mathbb{R}\omega \oplus \Lambda^{0,2}\,,$$

represented by the matrix

$$D = \begin{pmatrix} 0 & 0 & 0 \\ 0 & \lambda & 0 \\ 0 & 0 & 0 \end{pmatrix}.$$

Proof Under the inverse of the isomorphism (10.5) the symplectic form ω maps to

$$W = -f_1 \wedge g_1 - f_2 \wedge g_2 .$$

Therefore it suffices to show that

$$g(\mathcal{R}(W), u \wedge v) = \lambda g(W, u \wedge v) \quad \forall u, v \in T_p M .$$

The scalar product on the right-hand side can be identified with

$$\begin{aligned} g(W, u \wedge v) &= g(g_1 \wedge f_1, u \wedge v) + g(g_2 \wedge f_2, u \wedge v) \\ &= (\alpha_1 \wedge \beta_1)(u, v) + (\alpha_2 \wedge \beta_2)(u, v) \\ &= \omega(u, v) . \end{aligned}$$

The Einstein condition is equivalent to

$$r(v) = R(f_1, g_1)v + R(f_2, g_2)v = \lambda I(v) \quad \forall v \in T_p M .$$

It follows that

$$\begin{aligned} g(\mathcal{R}(W), u \wedge v) &= g(-R(f_1, g_1)v - R(f_2, g_2)v, u) \\ &= -\lambda g(I(v), u) \\ &= -\lambda \omega(v, u) \\ &= \lambda \omega(u, v) . \end{aligned}$$

This implies the claim. □

We can now finally prove the main theorem in this section.

Theorem 10.24 *Any Künneth–Einstein structure on a closed four-manifold M is Ricci-flat.*

Proof This follows from the Gauss–Bonnet Theorem and the Chern–Weil formulas for pseudo-Riemannian metrics of neutral signature [Ave-62, Law-91]. According to equation (4.3) in [Law-91], the Euler characteristic $\chi(M)$ and sig-

Four-manifolds

nature $\sigma(M)$ of M are related to the matrix D in Lemma 10.23 by

$$\frac{1}{4\pi^2}\lambda^2\text{vol}_g(M) = \frac{1}{4\pi^2}\int_M \text{tr}(D^2)\,\text{dvol}_g$$

$$= -\chi(M) - \frac{3}{2}\sigma(M)$$

$$= -\frac{1}{2}c_1^2(M),$$

where we used the Wu formula (10.2) in the last step. However, by Theorem 2.38 we have $c_1^2(M) = 0$. Therefore we conclude $\lambda = 0$, and hence the claim. □

10.6 T^2-bundles over T^2

In earlier chapters we have already seen several explicit examples of Künneth structures on specific T^2-bundles over T^2; compare Examples 5.22, 5.28 and 7.29. We also saw in Proposition 9.37 that nilmanifolds associated to the two-step nilpotent Lie group $Nil^3 \times \mathbb{R}$ admit hypersymplectic and hence Künneth structures. According to Theorem 9.25, these nilmanifolds are principal T^2-bundles over T^2. We now want to discuss all orientable T^2-bundles over T^2 more systematically.

Orientable smooth, not necessarily principal, T^2-bundles over T^2 were classified by Sakamoto–Fukuhara [SF-83]; see also [Ue-88, Ue-90]. Using this classification, Geiges [Gei-92] proved that all such bundles admit symplectic structures. Ideally, we would like to generalise this to prove a general existence theorem for Künneth structures. We do not quite achieve this goal here. We will prove existence of at least one Lagrangian foliation in all cases, and existence of Künneth structures in most, but not all, cases.

Every oriented T^2-bundle over T^2 can be constructed topologically by a two-step procedure. This construction depends on a pair of commuting matrices $A, B \in SL_2(\mathbb{Z})$ and a pair of integers (m, n). In the first step we construct a flat T^2-bundle over T^2, denoted by $M = \{A, B, (0, 0)\}$, by suspending the representation

$$\rho \colon \pi_1(T^2) \longrightarrow SL_2(\mathbb{Z})$$

sending the two generators of the fundamental group of the base to A and B respectively. Here, as before, $SL_2(\mathbb{Z})$ acts on T^2 by area-preserving linear diffeomorphisms. Because the action is area-preserving, it is clear that the resulting bundle will have a symplectic structure. If A and B preserve a Künneth structure on T^2, then the resulting flat bundle even has a Künneth structure.

In the second step, the bundle $M = \{A, B, (m, n)\}$ is constructed by cutting out a tubular neighbourhood $T^2 \times D^2$ of a fibre in the above flat bundle $M = \{A, B, (0, 0)\}$ and gluing it back in with the identification map

$$T^2 \times \partial D^2 \longrightarrow T^2 \times \partial D^2 ,$$

$$\left(\begin{bmatrix} s \\ t \end{bmatrix}, \theta \right) \longmapsto \left(\begin{bmatrix} s + m\theta/2\pi \\ t + n\theta/2\pi \end{bmatrix}, \theta \right).$$

It is not at all clear that the symplectic structure on $M = \{A, B, (0, 0)\}$ will survive under this surgery. For Künneth structures, checking survival under surgery seems very difficult.

Geiges's proof [Gei-92] that all these bundles are in fact symplectic relies on the work of Ue [Ue-90] showing that all orientable T^2-bundles over T^2 carry locally homogeneous geometric structures in the sense of Thurston [Thu-97]. We briefly digress to introduce this notion.

Definition 10.25 A *geometry* is a pair (X, G_X) where X is a simply connected manifold and G_X is a Lie group acting on the left on X such that:

(i) G_X acts transitively,
(ii) X admits a G_X-invariant Riemannian metric, equivalently the point stabilisers are compact,
(iii) G_X contains a discrete subgroup Γ such that $\Gamma \backslash X$ has finite volume.

A smooth manifold M obtained as a quotient $\Gamma \backslash X$, for a discrete subgroup $\Gamma \subset G$, is said to carry the geometry (X, G_X), or to be a *geometric manifold* for this geometry.

The geometries in small dimensions have been classified, and in dimensions ≤ 3 the classification is well known; cf. [Thu-97]. In dimension 4 the classification is due to Filipkiewicz (unpublished). We refer the reader to the account by Wall [Wall-85, Wall-86].

The simplest example of a geometry consists of a simply connected Lie group G acting on itself by left translation. The first and second conditions in the definition are always satisfied, and the third one is equivalent to the existence of a lattice of finite co-volume. This is a slight weakening of the notion of co-compactness. This example can be generalised as follows.

Example 10.26 Let G be a simply connected Lie group admitting a uniform lattice. Fix a left-invariant Riemannian metric g on G. Then the isometry group $\mathrm{Isom}(G, g)$, acting on G from the left, contains G as a subgroup and thus acts transitively on G. The pair $(G, \mathrm{Isom}(G))$ is a geometry.

If G is nilpotent, then the geometric manifolds with this geometry are the infra-nilmanifolds of G.

It turns out (see [Ue-90]), that all orientable T^2-bundles over T^2 are geometric manifolds for a geometry $(G, \text{Isom}(G))$, where G is one of the following four Lie groups

$$\mathbb{R}^4, \quad Nil^3 \times \mathbb{R}, \quad Nil^4, \quad Sol^3 \times \mathbb{R}$$

and $\text{Isom}(G)$ is the group of isometries of a left-invariant metric on G. We will usually suppress mentioning the isometry group, and simply speak of manifolds with the geometry G, the left-invariant metric being understood. Generators for the discrete subgroup $\Gamma \subset \text{Isom}(G)$ for each bundle $\Gamma \backslash G$ can be found in [Ue-88, Ue-90, Gei-92].

Theorem 10.27 *Let M be an orientable T^2-bundle over T^2. Then every cohomology class $a \in H^2(M; \mathbb{R})$ with $a^2 \neq 0$ can be represented by a symplectic form on M. Moreover, we can always choose the symplectic representative of the cohomology class a such that it admits a Lagrangian foliation.*

At least for the following bundles M we can choose the symplectic representative of a such that it admits a Künneth structure:

(i) *among the bundles of geometric type $\mathbb{R}^4$: the trivial bundle T^4 and the bundles with structure $\{-I, I, (-1, 0)\}$ and $\{-I, I, (0, 0)\}$,*
(ii) *the bundles of geometric type $Nil^3 \times \mathbb{R}$,*
(iii) *the bundles of geometric type $Sol^3 \times \mathbb{R}$.*

Table 10.1 summarises what we know about T^2-bundles over T^2. A check mark indicates that we have found Künneth structures, in fact for symplectic forms in all possible cohomology classes. A question mark indicates that we have found a Lagrangian foliation, but we do not know whether it admits a complementary Lagrangian foliation.

In the second column of Table 10.1, the integer $b_1(M)$ denotes the first Betti number of M. According to [Hil-02, p. 170], if M is a manifold of geometric type $Nil^3 \times \mathbb{R}$ or Nil^4, then

$$b_1(M) \leq 3 \quad \text{or} \quad 2,$$

respectively, with equality if and only if M is a nilmanifold. Therefore, cases (b) and (d) are nilmanifolds for $Nil^3 \times \mathbb{R}$ and Nil^4, whereas cases (e) and (f) are genuine infra-nilmanifolds for $Nil^3 \times \mathbb{R}$.

	$b_1(M)$	Structure of M	Geometric type	Künneth structure
(a)	4	$\{I, I, (0,0)\} = T^4$	$\mathbb{R}^4$	✓
(b)	3	$\{I, I, (m,n)\}, (m,n) \neq (0,0)$	$Nil^3 \times \mathbb{R}$	✓
(c)	2	$\left\{ \begin{pmatrix} 0 & -1 \\ 1 & -1 \end{pmatrix}, I, (0,0) \right\}$	$\mathbb{R}^4$	?
		$\left\{ \begin{pmatrix} 0 & -1 \\ 1 & -1 \end{pmatrix}, I, (-1,0) \right\}$	$\mathbb{R}^4$	?
		$\left\{ \begin{pmatrix} 0 & -1 \\ 1 & 0 \end{pmatrix}, I, (0,0) \right\}$	$\mathbb{R}^4$	?
		$\left\{ \begin{pmatrix} 0 & -1 \\ 1 & 0 \end{pmatrix}, I, (-1,0) \right\}$	$\mathbb{R}^4$	?
		$\left\{ \begin{pmatrix} 1 & -1 \\ 1 & 0 \end{pmatrix}, I, (0,0) \right\}$	$\mathbb{R}^4$	?
		$\{-I, I, (0,0)\}$	$\mathbb{R}^4$	✓
		$\{-I, I, (-1,0)\}$	$\mathbb{R}^4$	✓
(d)	2	$\left\{ \begin{pmatrix} 1 & \lambda \\ 0 & 1 \end{pmatrix}, I, (m,n) \right\}, \lambda \neq 0, n \neq 0$	Nil^4	?
(e)	2	$\left\{ \begin{pmatrix} -1 & \lambda \\ 0 & -1 \end{pmatrix}, I, (m,n) \right\}, \lambda \neq 0$	$Nil^3 \times \mathbb{R}$	✓
(f)	2	$\left\{ \begin{pmatrix} 1 & \lambda \\ 0 & 1 \end{pmatrix}, -I, (m,n) \right\}, \lambda \neq 0$	$Nil^3 \times \mathbb{R}$	✓
(g)	2	$\{C, I, (m,n)\}, |\text{tr } C| \geq 3, C \in SL_2(\mathbb{Z})$	$Sol^3 \times \mathbb{R}$	✓
(h)	2	$\{C, -I, (m,n)\}, \text{tr } C \geq 3, C \in SL_2(\mathbb{Z})$	$Sol^3 \times \mathbb{R}$	✓

Table 10.1 *Orientable T^2-bundles over T^2*

We frequently use the following lemma to define Lagrangian foliations.

Lemma 10.28   *Let (M, ω) be a symplectic four-manifold and $\eta \in \Omega^2(M)$ a two-form with the following properties:*

- *η is nowhere zero,*
- *$\eta \wedge \eta = 0$,*
- *$d\eta = 0$,*
- *$\omega \wedge \eta = 0$.*

Then

$$\ker \eta = \{X \in TM \mid i_X \eta = 0\}$$

is tangent to a Lagrangian foliation on M.

Proof   Considering the normal form of the two-form η_p on the tangent space $T_p M$ for each $p \in M$, the first two conditions imply that

$$\eta_p = \alpha_1 \wedge \alpha_2$$

for certain linearly independent one-forms $\alpha_1, \alpha_2 \in T_p^* M$. In particular, $\ker \eta$ is a subbundle of TM of rank 2.

Let X, Y be sections of $\ker \eta$. Then by equation (4.2)

$$0 = d\eta(X, Y, Z) = -\eta([X, Y], Z)$$

for all vector fields Z on M, hence $[X, Y]$ is a section of $\ker \eta$. Thus $\ker \eta$ is integrable and tangent to a foliation on M. Furthermore, if vectors $V, W \in TM$ are chosen so that $\eta(V, W) \neq 0$, then

$$0 = (\omega \wedge \eta)(X, Y, V, W) = \omega(X, Y)\eta(V, W)$$

implies that $\omega(X, Y) = 0$. It follows that $\ker \eta$ is Lagrangian. $\square$

We now prove Theorem 10.27.

Proof   We identify each of the four Lie groups G as a manifold with $\mathbb{R}^4$ with coordinates (x, y, z, t) and write $M = \Gamma \backslash \mathbb{R}^4$ with the discrete subgroup $\Gamma \subset \text{Isom}(G)$, where $\Gamma \cong \pi_1(M)$. The statement that every class $a \in H^2(M; \mathbb{R})$ with $a^2 \neq 0$ can be represented by a symplectic form on M was proved in [Gei-92].

We now go through the cases in Table 10.1.

(a) The case of the four-torus is trivial, of course, if one only wants a single Künneth structure. Nevertheless, we go through the proof as a warmup for the more complicated cases, showing that indeed all possible cohomology classes occur for the symplectic forms of Künneth structures.

To describe the generators of Γ we say that the point (x, y, z, t) goes to:

$$(x + 1, y, z, t), \quad (x, y + 1, z, t), \quad (x, y, z + 1, t) \quad \text{and} \quad (x, y, z, t + 1).$$

The two-forms

$$dx \wedge dy, \quad dx \wedge dz, \quad dx \wedge dt, \quad dy \wedge dz, \quad dy \wedge dt, \quad dz \wedge dt$$

are invariant under Γ and hence define forms on the quotient $M = \Gamma \backslash \mathbb{R}^4$. The cohomology classes defined by these closed forms span $H^2(M; \mathbb{R}) \cong \mathbb{R}^6$. A linear combination ω of these two-forms is symplectic if and only if the class $a = [\omega]$ has non-zero square. Therefore every class with non-zero square can be represented by a symplectic form ω. We can perform a linear change of coordinates such that ω takes the form

$$\omega = dx \wedge dy + dz \wedge dt.$$

This changes the generators of Γ, but the basis vectors $e_1, \dots, e_4$ are still invariant. Let

$$\mathcal{F} = \mathrm{span}\{e_1, e_3\}, \quad \mathcal{G} = \mathrm{span}\{e_2, e_4\}.$$

Then this defines a Γ-invariant Künneth structure on $\mathbb{R}^4$ and hence a Künneth structure on the quotient M.

(b) This is the case of nilmanifolds for $Nil^3 \times \mathbb{R}$, and the result follows from Proposition 9.37.

(c) This is the case of hyperelliptic complex surfaces. In this and all the remaining cases we have $H^2(M; \mathbb{R}) \cong \mathbb{R}^2$. To define a Lagrangian foliation we have to consider each case separately. We use the numbering $M_1, \dots, M_7$ from [Ue-88], where the generators of Γ for M_1, M_2, M_3 can also be found. The manifolds $M_4, \dots, M_7$ are of the form $\{A, I, (0, 0)\}$ with a matrix $A \in SL_2(\mathbb{Z})$. The four generators of Γ are given by mapping $\left(\begin{pmatrix} x \\ y \end{pmatrix}, \begin{pmatrix} z \\ t \end{pmatrix} \right)$ to

$$\left(\begin{pmatrix} x+1 \\ y \end{pmatrix}, A^{-1} \begin{pmatrix} z \\ t \end{pmatrix} \right), \quad \left(\begin{pmatrix} x \\ y+1 \end{pmatrix}, \begin{pmatrix} z \\ t \end{pmatrix} \right),$$

$$\left(\begin{pmatrix} x \\ y \end{pmatrix}, \begin{pmatrix} z+1 \\ t \end{pmatrix} \right), \quad \left(\begin{pmatrix} x \\ y \end{pmatrix}, \begin{pmatrix} z \\ t+1 \end{pmatrix} \right).$$

$M_4 = \{-I, I, (0, 0)\}$: An example of a Künneth structure has been constructed in Example 7.29. Note that the Künneth structure given there is not orientable, since the Lagrangian foliations have non-orientable leaves.

In the general case, the generators of Γ are

$$(x+1, y, -z, -t), \quad (x, y+1, z, t), \quad (x, y, z+1, t) \quad \text{and} \quad (x, y, z, -t).$$

The cohomology $H^2(M_4; \mathbb{R})$ is generated by the invariant forms $dx \wedge dy$ and $dz \wedge dt$. Let ω be a symplectic linear combination. Then

$$\mathcal{F} = \text{span}\{e_1, e_3\}, \quad \mathcal{G} = \text{span}\{e_2, e_4\}$$

define a Künneth structure on M_4.

$M_1 = \{-I, I, (-1, 0)\}$, where the generators of Γ are

$$\left(x, y + 1, z - \tfrac{1}{2}, t\right), \quad (x, y, z + 1, t), \quad (x, y, z, t + 1) \quad \text{and} \quad (x + 1, y, -z, -t).$$

The cohomology $H^2(M_1; \mathbb{R})$ is generated by the invariant forms $dx \wedge dy$ and $dz \wedge dt$. Let ω be a symplectic linear combination. Then

$$\mathcal{F} = \text{span}\{e_1, e_3\}, \quad \mathcal{G} = \text{span}\{e_2, e_4\}$$

define a Künneth structure on M_1.

$$M_5 = \left\{ \begin{pmatrix} 0 & -1 \\ 1 & -1 \end{pmatrix}, I, (0, 0) \right\},$$

where the generators of Γ are given by

$$(x + 1, y, t - z, -z), \quad (x, y + 1, z, t), \quad (x, y, z + 1, t) \quad \text{and} \quad (x, y, z, t + 1).$$

The cohomology $H^2(M_5; \mathbb{R})$ is generated by the invariant forms $dx \wedge dy$ and $dz \wedge dt$. Let

$$\omega = \alpha dx \wedge dy + \beta dz \wedge dt$$

be a symplectic linear combination. Define a nowhere zero two-form η by

$$\eta = g(x) dx \wedge dz + f(x) dx \wedge dt,$$

where

$$f(x) = \cos\left(\tfrac{4}{3}\pi x\right), \quad g(x) = f(x - 1).$$

The function f satisfies

$$f(x - 1) + f(x) + f(x + 1) = 0.$$

We check that η is invariant under Γ. This is non-trivial only for the first generator. Let ϕ be the map corresponding to this generator. Then in the point

$p = (x, y, z, t) \in \mathbb{R}^4$ we have

$$
\begin{aligned}
(\phi^* \eta)_p &= g(x+1)\phi^* dx \wedge \phi^* dz + f(x+1)\phi^* dx \wedge \phi^* dt \\
&= f(x)dx \wedge (dt - dz) + f(x+1)dx \wedge (-dz) \\
&= -(f(x) + f(x+1))dx \wedge dz + f(x)dx \wedge dt \\
&= f(x-1)dx \wedge dz + f(x)dx \wedge dt \\
&= \eta_p.
\end{aligned}
$$

The two-form η is closed and we have $\eta \wedge \eta = 0 = \omega \wedge \eta$, hence η defines a Lagrangian foliation on M_5.

$$
M_2 = \left\{ \begin{pmatrix} 0 & -1 \\ 1 & -1 \end{pmatrix}, I, (-1, 0) \right\},
$$

where the generators of Γ are given by

$$(x, y+1, z, t), \quad (x, y, z+1, t), \quad (x, y, z, t+1) \quad \text{and} \quad (x+1, t, y, z).$$

The two-forms

$$dx \wedge dy + dx \wedge dz + dx \wedge dt \quad \text{and} \quad dy \wedge dz + dz \wedge dt + dt \wedge dy$$

are invariant under Γ and generate $H^2(M_2; \mathbb{R})$. Let ω be a linear combination of these forms. It is symplectic if and only if the coefficients are both non-zero. Consider the two-form

$$\eta = f(x)dx \wedge dy + g(x)dx \wedge dz + h(x)dx \wedge dt,$$

where the functions are given by

$$f(x) = \cos\left(\tfrac{4}{3}\pi x\right), \quad g(x) = f(x-1) \quad \text{and} \quad h(x) = f(x+1).$$

Note that $f(x) + g(x) + h(x) = 0$, so $\omega \wedge \eta = 0$. Moreover, η is closed, $\eta \wedge \eta = 0$ and a similar calculation as for M_5 shows that η is invariant under the action of Γ, hence it defines a Lagrangian foliation on M_2.

$$
M_6 = \left\{ \begin{pmatrix} 0 & -1 \\ 1 & 0 \end{pmatrix}, I, (0, 0) \right\},
$$

where the generators of Γ are

$$(x+1, y, t, -z), \quad (x, y+1, z, t), \quad (x, y, z+1, t) \quad \text{and} \quad (x, y, z, t+1).$$

The second cohomology of M_6 is generated by the invariant forms $dx \wedge dy$ and $dz \wedge dt$. Let ω be a symplectic linear combination and consider the two-form

$$\eta = g(x)dx \wedge dz + f(x)dx \wedge dt$$

with functions

$$f(x) = \sin\left(\tfrac{\pi}{2}x\right) \quad \text{and} \quad g(x) = -\cos\left(\tfrac{\pi}{2}x\right).$$

Again, η is closed, invariant under Γ and $\eta \wedge \eta = 0 = \omega \wedge \eta$, hence it defines a Lagrangian foliation on M_6.

$$M_3 = \left\{ \begin{pmatrix} 0 & -1 \\ 1 & 0 \end{pmatrix}, I, (-1,0) \right\},$$

where the generators of Γ are

$$\left(x, y+1, z - \tfrac{1}{2}, t - \tfrac{1}{2}\right), \quad (x, y, z+1, t), \quad (x, y, z, t+1) \quad \text{and} \quad (x+1, y, -t, z).$$

The Γ-invariant two-forms $dx \wedge dy$ and $dz \wedge dt$ generate $H^2(M_3; \mathbb{R})$. Let ω be a symplectic linear combination and define a two-form η by

$$\eta = g(x)dx \wedge dz + f(x)dx \wedge dt,$$

where the functions f, g are given by

$$f(x) = \sin\left(\tfrac{\pi}{2}x\right) \quad \text{and} \quad g(x) = \cos\left(\tfrac{\pi}{2}x\right).$$

Then η is invariant under Γ, $d\eta = 0$ and $\eta \wedge \eta = 0 = \omega \wedge \eta$, hence it defines a Lagrangian foliation on M_3.

$$M_7 = \left\{ \begin{pmatrix} 1 & -1 \\ 1 & 0 \end{pmatrix}, I, (0,0) \right\},$$

where the generators of Γ are

$$(x+1, y, t, t-z), \quad (x, y+1, z, t), \quad (x, y, z+1, t) \quad \text{and} \quad (x, y, z, t+1),$$

and the second cohomology of M_7 is generated by the two-forms $dx \wedge dy$ and $dz \wedge dt$. Let ω be a symplectic linear combination and consider the two-form

$$\eta = g(x)dx \wedge dz + f(x)dx \wedge dt,$$

where the functions are

$$f(x) = \cos\left(\tfrac{2}{3}\pi x\right) \quad \text{and} \quad g(x) = -f(x+1).$$

Again, η is invariant under Γ, closed and satisfies $\eta \wedge \eta = 0 = \omega \wedge \eta$. Hence we get a Lagrangian foliation on M_7.

(d) This is the case of nilmanifolds for Nil^4, and the claims follow from Proposition 9.40.

(e), (f) These are the infra-nilmanifolds for $Nil^3 \times \mathbb{R}$. Although we have lots of left-invariant Künneth structures, one has to check that they descend to the quotients by certain automorphisms.

Generators of Γ are

$$(x, y, z + \gamma_1, t), \quad (x + \alpha_1, y, z + \alpha_1 y, t),$$

$$(x + \alpha_2, y, z + \alpha_2 y + \gamma_2, t + 1) \quad \text{and} \quad (-x, y + 1, -z + \gamma_3, t)$$

in case (e); for case (f) modify the last two generators to

$$(x + \alpha_3, y + 1, z + \alpha_3 y + \gamma_4, t) \quad \text{and} \quad (-x, y, -z, t + 1).$$

Here α_i, γ_j are certain real constants. The cohomology $H^2(M; \mathbb{R})$ is generated by the invariant forms $dy \wedge dt$ and $dx \wedge dz - x dx \wedge dy$. Let ω be a symplectic linear combination. The foliations

$$\mathcal{F} = \text{span}\{e_1, e_4\}, \quad \mathcal{G} = \text{span}\{e_2, e_3\}$$

are invariant, Lagrangian and complementary, and hence define a Künneth structure on M.

(g), (h) These cases were essentially done in Section 9.7 of Chapter 9. Generators of Γ are

$$(x + \alpha_1, y + \beta_1, z, t), \quad (x + \alpha_2, y + \beta_2, z, t)$$

$$(x + \alpha_3, y + \beta_3, z + 1, t) \quad \text{and} \quad (\epsilon e^\delta x + \alpha_4, \epsilon e^{-\delta} y + \beta_4, z, t + \delta)$$

in case (g); for (h) modify the third generator to

$$(-x + \alpha_3, -y + \beta_3, z + 1, t).$$

Here the cohomology $H^2(M; \mathbb{R})$ is generated by the invariant forms $dx \wedge dy$ and $dz \wedge dt$. Let ω be a symplectic linear combination. The foliations

$$\mathcal{F} = \text{span}\{e_1, e_3\}, \quad \mathcal{G} = \text{span}\{e_2, e_4\}$$

are invariant and Lagrangian. We get a Künneth structure on the quotient M.

This finally completes the proof of Theorem 10.27. $\qquad\qquad\square$

10.7 Open Problems

There are several open problems about the classification of Künneth manifolds that arise naturally from the results discussed in this chapter.

First of all, there is the well-known problem of classifying closed symplectic Calabi–Yau four-manifolds. This problem is discussed, for example, in [Bau-08, Li-06, FV-13]. In view of Theorem 10.13 the problem splits naturally into two cases, according to whether the first Betti number vanishes or not.

It seems quite reasonable to conjecture that a symplectic Calabi–Yau with $b_1 = 0$ must be simply connected, and, even stronger, diffeomorphic to the $K3$ manifold. This would be the optimal strengthening of Corollaries 10.14 and 10.15. If $b_1 > 0$, then one might conjecture that any SCY manifold is in fact a T^2-bundle over T^2, or at least has such a bundle as a finite covering. These conjectures amount to saying that the currently known SCY four-manifolds exhaust this class.

Second of all, there is the problem of deciding, for every SCY four-manifold, whether it admits a Künneth structure or not. In view of the results proved here, the most interesting problem is certainly the following.

Problem 10.29 *Does the $K3$ manifold admit a Künneth structure?*

For manifolds with positive first Betti number we have the following.

Problem 10.30 *Do any nilmanifolds for Nil^4 admit a Künneth structure?*

We have shown in Proposition 9.40 that these manifolds have symplectic forms admitting Lagrangian foliations, and that they do not admit left-invariant Künneth structures. So if there are any Künneth structures at all, they must be constructed in a much more complicated way.

Finally, in arbitrary dimensions that are multiples of 4, we have the following.

Problem 10.31 *Do there exist closed manifolds with non-zero Euler characteristic that admit a Künneth structure?*

An affirmative answer to Problem 10.29 would of course solve this, and would also give high-dimensional examples by taking products. Recall that by Theorem 5.7 the Euler characteristics of almost Künneth manifolds of dimension $4k + 2$ vanish. This is no longer true in dimension $4k$, but all the examples we have are of non-integrable structures.

Notes for Chapter 10

1. Proposition 10.11 is due to Kamada [Kam-02]. The proof we have given here is due to Bande–Kotschick [BK-08].

2. Corollary 10.15 was proved by Morgan–Szabó [MoSz-97] long before Theorem 10.13.

3. The conclusion of Proposition 10.17 has appeared in papers of Friedl and Vidussi [FV-13, FV-15], always under some additional assumption (residual finiteness and the vanishing of the first L^2-Betti number, respectively). The argument we have given here, going back to [JK-93, Theorem 4], shows that none of these additional assumptions are necessary.

4. The five-dimensional Thurston geometries have been classified by Geng [Gen-16].

Bibliography

[AMT-09]   D. V. Alekseevsky, C. Medori and A. Tomassini, Homogeneous para-Kähler Einstein manifolds, *Russ. Math. Surv.* **64** (2009), 1–43.

[And-06]   A. Andrada, Hypersymplectic Lie algebras, *J. Geom. Phys.* **56** (2006), 2039–2067.

[AD-06]   A. Andrada and I. G. Dotti, Double products and hypersymplectic structures on $\mathbb{R}^{4n}$, *Commun. Math. Phys.* **262** (2006), 1–16.

[Ave-62]   A. Avez, Formule de Gauss-Bonnet-Chern en métrique de signature quelconque, *C. R. Acad. Sci. Paris* **255** (1962), 2049–2051.

[BK-06]   G. Bande and D. Kotschick, The geometry of symplectic pairs, *Trans. Amer. Math. Soc.* **358** (2006), 1643–1655.

[BK-08]   G. Bande and D. Kotschick, The geometry of recursion operators, *Commun. Math. Phys.* **280** (2008), 737–749.

[BaRR-16]   B. Banos, V. N. Roubtsov and I. Roulstone, Monge–Ampère structures and the geometry of incompressible flows, *J. Phys. A* **49** (2016), no. 24, 244003, 17 pp.

[BPV-84]   W. Barth, C. Peters and A. Van de Ven, *Compact Complex Surfaces*, Springer-Verlag, Berlin 1984.

[Bau-08]   S. Bauer, Almost complex 4-manifolds with vanishing first Chern class, *J. Differ. Geom.* **79** (2008), 25–32.

[BC-16]   O. Baues and V. Cortés, *Symplectic Lie Groups*, Astérisque **379**, Société Mathématique de France, Marseille 2016.

[BM-12]   G. Bazzoni and V. Muñoz, Classification of minimal algebras over any field up to dimension 6, *Trans. Amer. Math. Soc.* **364** (2012), no. 2, 1007–1028.

[Bej-93]   C.-L. Bejan, The existence problem of hyperbolic structures on vector bundles, *Publ. Inst. Math. (Beograd) (N.S.)* **53** (1993), no. 67, 133–138.

[Bej-94]   C.-L. Bejan, Some examples of manifolds with hyperbolic structures, *Rend. Mat. Appl.*, Ser. VII, **14** (1994), no. 4, 557–565.

[BFL-92]   Y. Benoist, P. Foulon and F. Labourie, Flots d'Anosov à distributions stable et instable différentiables, *J. Amer. Math. Soc.* **5** (1992), 33–74.

[BL-93]   Y. Benoist and F. Labourie, Sur les difféomorphismes d'Anosov affines à feuilletages stable et instable différentiables, *Invent. Math.* **111** (1993), 285–308.

[BCS-13]   M. Bestvina, T. Church and J. Souto, Some groups of mapping classes not realized by diffeomorphisms, *Comment. Math. Helv.* **88** (2013), no. 1, 205–220.

[BiRR-17]   R. Bielawski, N. M. Romão and M. Röser, The Nahm–Schmid equations and hypersymplectic geometry, *Q. J. Math.* **69** (2018), no. 4, 1253–1286.

[BN-84]   G. S. Birman and K. Nomizu, The Gauss-Bonnet theorem for 2-dimensional spacetimes, *Michigan Math. J.* **31** (1984), 77–81.

[Bow-11]   J. Bowden, Flat structures on surface bundles, *Algebraic Geom. Topol.* **11** (2011), no. 4, 2207–2235.

[Boy-89]   N. B. Boyom, Varietes symplectiques affines, *Manuscripta Math.* **64** (1989), 1–33.

[Boy-95]   N. B. Boyom, Structures localement plates dans certaines variétés symplectiques, *Math. Scand.* **76** (1995), 61–84.

[Bry-01]   R. L. Bryant, Bochner-Kähler metrics, *J. Amer. Math. Soc.* **14** (2001), 623–715.

[BILW-05]   M. Burger, A. Iozzi, F. Labourie and A. Wienhard, Maximal representations of surface groups: symplectic Anosov structures, *Pure Appl. Math. Q.* **1** (2005), no. 3, Special Issue: In memory of Armand Borel, Part 2, 543–590.

[Car-94]   Y. Carrière, Autour de la conjecture de L. Markus sur les variétés affines, *Invent. Math.* **95** (1989), 615–628.

[CMMS-04]   V. Cortés, C. Mayer, T. Mohaupt and F. Saueressig, Special geometry of Euclidean supersymmetry. I. Vector multiplets, *J. High Energy Phys.* **2004**, no. 3, 028, 73 pp.

[CFG-96]   V. Cruceanu, P. Fortuny and P. M. Gadea, A survey on paracomplex geometry, *Rocky Mountain J. Math.* **26** (1996), 83–115.

[DG-91]   G. D'Ambra and M. Gromov, Lectures on transformation groups: geometry and dynamics, in *Surveys in Differential Geometry* (Cambridge, MA, 1990), 19–111, Lehigh University, Bethlehem, PA, 1991.

[DJS-08]   A. S. Dancer, H. R. Jørgensen and A. F. Swann, Metric geometries over the split quaternions, *Rend. Sem. Mat. Univ. Pol. Torino* **63** (2005), 119–139.

[DS-05]   A. S. Dancer and A. F. Swann, Hypersymplectic manifolds, in *Recent Developments in Pseudo-Riemannian Geometry*, 97–111, ESI Lect. Math. Phys., European Mathematical Society Press, Zürich 2008.

[Daz-81]   P. Dazord, Sur la géométrie des sous-fibrés et des feuilletages Lagrangiens, *Ann. scient. Éc. Norm. Sup.* **14** (1981), 465–480.

[Daz-85]   P. Dazord, Erratum to [Daz-81], *Ann. scient. Éc. Norm. Sup.* **18** (1985), 685.

[Deh-12]   M. Dehn, Über unendliche diskontinuierliche Gruppen, *Math. Ann.* **71** (1912), 116–144.

[DW-59]   A. Dold and H. Whitney, Classification of oriented sphere bundles over a 4-complex, *Ann. of Math.* **69** (1959), 667–677.

[Don-86]   S. K. Donaldson, Connections, cohomology and the intersection forms of 4-manifolds, *J. Differ. Geom.* **24** (1986), 275–341.

[DK-90]   S. K. Donaldson and P. B. Kronheimer, *The Geometry of Four-manifolds*, Oxford University Press, Oxford 1990.

[Dui-80]   J. J. Duistermaat, On global action-angle coordinates, *Commun. Pure Appl. Math.* **33** (1980), 687–706.

[ES-01]   F. Etayo Gordejuela and R. Santamaría, The canonical connection of a bi-Lagrangian manifold, *J. Phys. A* **34** (2001), 981–987.

[EST-06]   F. Etayo, R. Santamaría and U. R. Trías, The geometry of a bi-Lagrangian manifold, *Differ. Geom. Appl.* **24** (2006), 33–59.

[FPPS-04]   A. Fino, H. Pedersen, Y.-S. Poon and M. W. Sørensen, Neutral Calabi–Yau structures on Kodaira manifolds, *Commun. Math. Phys.* **248** (2004), 255–268.

[FY-13]   M. Forger and S. Z. Yepes, Lagrangian distributions and connections in multisymplectic and polysymplectic geometry, *Differ. Geom. Appl.* **31** (2013), no. 6, 775–807.

[FV-13]   S. Friedl and S. Vidussi, On the topology of symplectic Calabi–Yau 4-manifolds, *J. Topol.* **6** (2013), 945–954.

[FV-15]   S. Friedl and S. Vidussi, Thompson's group F is not SCY, *Groups Geom. Dyn.* **9** (2015), 325–329.

[Gei-92]   H. Geiges, Symplectic structures on T^2-bundles over T^2, *Duke Math. J.* **67** (1992), 539–555.

[Gei-96]   H. Geiges, Symplectic couples on 4-manifolds, *Duke Math. J.* **85** (1996), 701–711.

[Gen-16]   A. Geng, 5-dimensional geometries I: the general classification, Preprint arxiv:1605.07545v2 [math.GT] 8 Jun 2016.

[GL-16]   A. Gogolev and J.-F. Lafont, Aspherical products which do not support Anosov diffeomorphisms, *Ann. Henri Poincaré* **17** (2016), no. 11, 3005–3026.

[GOV-97]   V. V. Gorbatsevich, A. L. Onishchik and E. B. Vinberg, *Foundations of Lie Theory and Lie Transformation Groups*, Springer-Verlag, Berlin 1997.

[Gro-88]   M. Gromov, Rigid transformations groups, in *Géométrie différentielle* (Paris, 1986), 65–139, Travaux en Cours, 33, Hermann, Paris, 1988.

[Gui-00]   V. Guillemin, Künneth geometry, Seminar talk at Harvard, 4 December 2000.

[Ham-04]   M. J. D. Hamilton, *Bi-Lagrangian Structures on Closed Manifolds*, Diplomarbeit, LMU München 2004.

[Ham-19]   M. J. D. Hamilton, Bi-Lagrangian structures on nilmanifolds, *J. Geom. Phys.* **140** (2019), 10–25.

[HK-21]   M. J. D. Hamilton and D. Kotschick, Lagrangian foliations and Anosov symplectomorphisms on Kähler manifolds, *Ergod. Theory Dyn. Syst.* **41** (2021), 3325–3335.

[HL-12]   F. R. Harvey and H. B. Lawson, Split special Lagrangian geometry, in *Metric and Differential Geometry: The Jeff Cheeger Anniversary Volume*, eds. X. Dai and X. Rong, 43–89, Prog. in Math. **297**, Springer, Basel 2012.

[Hes-80]   H. Hess, Connections on symplectic manifolds and geometric quantization, *Lect. Notes Math.* **836** (1980), 153–166.

[Hes-81]   H. Hess, On a geometric quantization scheme generalizing those of Kostant-Souriau and Czyz, *Lect. Notes Phys.* **139** (1981), 1–35.

[Hil-02]   J. A. Hillman, *Four-manifolds, Geometries and Knots*, Geometry & Topology Monographs 2002.

[Hi-54]   F. Hirzebruch, Some problems on differentiable and complex manifolds, *Ann. Math.* **60** (1954), 213–236; reprinted in [Hi-87].

[Hi-87]   F. Hirzebruch, *Gesammelte Abhandlungen*, Band I, Springer-Verlag, Berlin 1987.

[Hit-90]   N. J. Hitchin, Hypersymplectic quotients, in *La 'Mécanique analytique' de Lagrange et son héritage*, Supplemento al numero **124** (1990) degli Atti della Accademia delle Scienze di Torino, Classe de Scienze Fisiche, Matematiche e Naturali, 169–180.

[HDK-97]   Z. Hou, S. Deng and S. Kaneyuki, Dipolarizations in compact Lie algebras and homogeneous parakähler manifolds, *Tokyo J. Math.* **20** (1997), 381–388.

[HDKN-99]   Z. Hou, S. Deng, S. Kaneyuki and K. Nishiyama, Dipolarizations in semisimple Lie algebras and homogeneous parakähler manifolds, *J. Lie Theory* **9** (1999), 215–232.

[JK-93]   F. E. A. Johnson and D. Kotschick, On the signature and Euler characteristic of certain four-manifolds, *Math. Proc. Cambridge Philos. Soc.* **114** (1993), no. 3, 431–437.

[JW-00]   F. E. A. Johnson and J. P. Walton, Parallelizable manifolds and the fundamental group, *Mathematika* **47** (2000), nos. 1–2, 165–172.

[Kam-02]   H. Kamada, Self-dual Kähler metrics of neutral signature on complex surfaces, PhD thesis, Tohoku University, Sendai 2002; published as *Tohoku Math. Publ.* **24** (2002).

[Kan-88]   M. Kanai, Geodesic flows of negatively curved manifolds with smooth stable and unstable foliations, *Ergod. Theory Dyn. Syst.* **8** (1988), 215–239.

[KGM-04]   Yu. Khakimdjanov, M. Goze and A. Medina, Symplectic or contact structures on Lie groups, *Differ. Geom. Appl.* **21** (2004), 41–54.

[Kli-17]   B. Klingler, Chern's conjecture for special affine manifolds, *Ann. of Math.* **186** (2017), 69–95.

[Kna-96]   A. W. Knapp, *Lie Groups Beyond an Introduction*, Birkhäuser, Basel 1996.

[Kot-94]   D. Kotschick, Four-manifold invariants of finitely presentable groups, in *Topology, Geometry and Field Theory*, 89–99, World Scientific Publishing, River Edge, NJ 1994.

[Kot-97a]   D. Kotschick, The Seiberg–Witten invariants of symplectic four–manifolds, *Séminaire Bourbaki*, 48ème année, 1995–96, no. 812, Astérisque **241** (1997), 195–220.

[Kot-97b]   D. Kotschick, Orientations and geometrisations of compact complex surfaces, *Bull. Lond. Math. Soc.* **29** (1997), 145–149.

[Kot-13]   D. Kotschick, Updates on Hirzebruch's 1954 Problem List, Preprint arXiv:1305.4623v1 [math.HO] 20 May 2013.

[KM-05]   D. Kotschick and S. Morita, Signatures of foliated surface bundles and the symplectomorphism groups of surfaces, *Topology* **44** (2005), no. 1, 131–149.

[Law-91]   P. Law, Neutral Einstein metrics in four dimensions, *J. Math. Phys.* **32** (1991), no. 11, 3039–3042.

[Li-06]   T.-J. Li, Quaternionic bundles and Betti numbers of symplectic four-manifolds with Kodaira dimension zero, *Int. Math. Res. Not.* 2006, Article ID 37385, pages 1–28.

[Lib-52]   P. Libermann, Sur les structures presque paracomplexes, (French) *C. R. Acad. Sci. Paris* **234** (1952), 2517–2519.

[Lib-54]   P. Libermann, Sur le problème d'équivalence de certaines structures infinitésimales, (French) *Ann. Mat. Pura Appl.* **36** (1954), 27–120.

[LS-17]   B. Loustau and A. Sanders, Bi-Lagrangian structures and Teichmüller theory, Preprint arxiv:1708.09145v1 [math.DG] 30 Aug 2017.

[Malc-62]   A. I. Malcev, On a class of homogeneous spaces, *Am. Math. Soc. Transl.* **9** (Series 1) (1962), 276–307.

[Malf-00]   W. Malfait, Anosov diffeomorphisms on nilmanifolds of dimension at most six, *Geom. Dedicata* **79** (2000), 291–298.

[McS-95]   D. McDuff and D. Salamon, *Introduction to Symplectic Topology*, Clarendon Press, Oxford 1995.

[MS-74]   J. W. Milnor and J. D. Stasheff, *Characteristic Classes*, Ann. of Math. Studies **76**, Princeton University Press, Princeton, NJ 1974.

[Mor-96]   J. W. Morgan, *The Seiberg–Witten Equations and Applications to the Topology of Smooth Four-manifolds*, Princeton University Press, Princeton, NJ 1996.

[MoSz-97]   J. W. Morgan and Z. Szabó, Homotopy $K3$ surfaces and mod 2 Seiberg–Witten invariants, *Math. Res. Lett.* **4** (1997), no. 1, 17–21.

[Mos-65]   J. Moser, On the volume elements on a manifold, *Trans. Amer. Math. Soc.* **120** (1965), 286–294.

[Mou-03]   P. Mounoud, Dynamical properties of the space of Lorentzian metrics, *Comment. Math. Helv.* **78** (2003), 463–485.

[Nom-54]   K. Nomizu, On the cohomology of compact homogeneous spaces of nilpotent Lie groups, *Ann. of Math.* **59** (1954), 531–538.

[PS-61]   R. S. Palais and T. E. Stewart, Torus bundles over a torus, *Proc. Amer. Math. Soc.* **12** (1961), 26–29.

[Ras-48]   P. K. Raševskiǐ, The scalar field in a stratified space, (Russian) *Trudy Sem. Vektor. Tenzor. Analizu* **6** (1948) 225–248.

[Sal-01]   S. M. Salamon, Complex structures on nilpotent Lie algebras, *J. Pure Appl. Algebra* **157** (2001), 311–333.

[SF-83]   K. Sakamoto and S. Fukuhara, Classification of T^2-bundles over T^2, *Tokyo J. Math.* **6** (1983), 311–327.

[Ser-79]   J.-P. Serre, *A Course in Arithmetic*, Springer Verlag, New York 1979.

[Shub-69]   M. Shub, Endomorphisms of compact differentiable manifolds, *Am. J. Math.* **91** (1969), 175–199.

[Sma-67]   S. Smale, Differentiable dynamical systems, *Bull. Amer. Math. Soc.* **73** (1967), 747–817.

[Tab-93]   S. Tabachnikov, Geometry of Lagrangian and Legendrian 2-web, *Differ. Geom. Appl.* **3** (1993), 265–284.

[Tau-95]   C. H. Taubes, More constraints on symplectic forms from Seiberg–Witten invariants, *Math. Res. Lett.* **2** (1995), no. 1, 9–13.

[Thu-74]   W. P. Thurston, The theory of foliations of codimension greater than one, *Comment. Math. Helv.* **49** (1974), 214–231.

[Thu-76a]   W. P. Thurston, Existence of codimension-one foliations, *Ann. of Math.* **104** (1976), 249–268.

[Thu-76b]   W. P. Thurston, Some simple examples of symplectic manifolds, *Proc. Amer. Math. Soc.* **55** (1976), 467–468.

[Thu-97]   W. P. Thurston, *Three Dimensional Geometry and Topology*, Princeton University Press, Princeton, NJ 1997.

[Ue-88]   M. Ue, On the 4-dimensional Seifert fiberings with Euclidean base orbifolds, in *A Fête of Topology*, 471–523, Academic Press, Boston, MA 1988.

[Ue-90]   M. Ue, Geometric 4-manifolds in the sense of Thurston and Seifert 4-manifolds I, *J. Math. Soc. Japan* **42** (1990), 511–540.

[Vai-87]   I. Vaisman, *Symplectic Geometry and Secondary Characteristic Classes*, Progress in Math. **72**, Birkhäuser, Basel 1987.

[Vai-89]   I. Vaisman, Basics of Lagrangian foliations, *Publ. de l'Institut Math.* **33** (1989), no. 3, 559–575.

[Walc-05]   R. Walczak, Existence of symplectic structures on torus bundles over surfaces, *Ann. Glob. Anal. Geom.* **28** (2005), 211–231.

[Wall-85]   C. T. C. Wall, Geometries and geometric structures in real dimension 4 and complex dimension 2, *Lect. Notes Math.* **1167** (1985), 268–292.

[Wall-86]   C. T. C. Wall, Geometric structures on compact complex analytic surfaces, *Topology* **25** (1986), 119–153.

[Wei-71]   A. Weinstein, Symplectic manifolds and their Lagrangian submanifolds, *Adv. Math.* **6** (1971), 329–346.

[Wei-77]   A. Weinstein, *Lectures on Symplectic Manifolds*, CBMS Lecture Notes, Vol. **29**, American Mathematical Society, Providence, RI 1977.

Index